Build the S.M.A.R.T. Alarm Clock.

www.radioshack.com/DIT

PARTS

- ○ Arduino Yún microcontroller/computer board RadioShack® 276-357
- ○ 2.8" TFT Touch Shield, for Arduino RadioShack® 277-070/276-382
- ○ USB speaker such as RadioShack® 40-380
  (Make sure your speaker receives both power and audio over USB.)
- ○ MicroSD Card such as RadioShack® 55060858
  (Any size card will work; at least 15MB of space is required.)
- ○ USB cable, Micro-B to Standard-A RadioShack® 26-3260
- ○ AC adapter with USB, Enercell® 5V/3.6A RadioShack® 273-437

TOOLS

- ○ Computer with USB port, running Arduino IDE Version 1.5.6-r2 BETA
  free download from arduino.cc

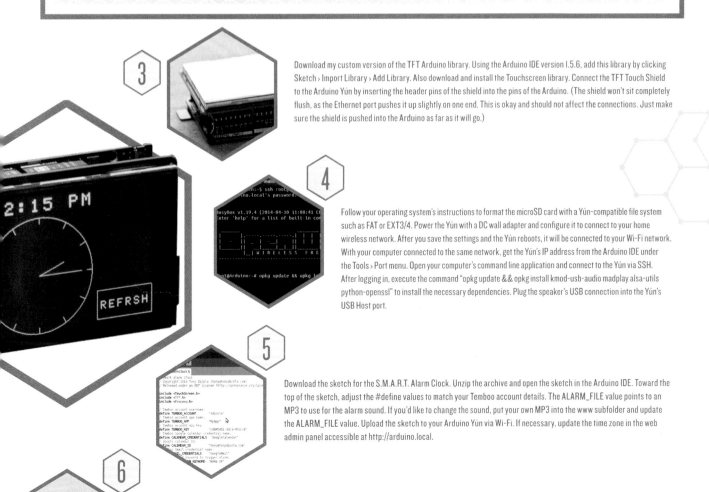

3 Download my custom version of the TFT Arduino library. Using the Arduino IDE version 1.5.6, add this library by clicking Sketch › Import Library › Add Library. Also download and install the Touchscreen library. Connect the TFT Touch Shield to the Arduino Yún by inserting the header pins of the shield into the pins of the Arduino. (The shield won't sit completely flush, as the Ethernet port pushes it up slightly on one end. This is okay and should not affect the connections. Just make sure the shield is pushed into the Arduino as far as it will go.)

4 Follow your operating system's instructions to format the microSD card with a Yún-compatible file system such as FAT or EXT3/4. Power the Yún with a DC wall adapter and configure it to connect to your home wireless network. After you save the settings and the Yún reboots, it will be connected to your Wi-Fi network. With your computer connected to the same network, get the Yún's IP address from the Arduino IDE under the Tools › Port menu. Open your computer's command line application and connect to the Yún via SSH. After logging in, execute the command "opkg update && opkg install kmod-usb-audio madplay alsa-utils python-openssl" to install the necessary dependencies. Plug the speaker's USB connection into the Yún's USB Host port.

5 Download the sketch for the S.M.A.R.T. Alarm Clock. Unzip the archive and open the sketch in the Arduino IDE. Toward the top of the sketch, adjust the #define values to match your Temboo account details. The ALARM_FILE value points to an MP3 to use for the alarm sound. If you'd like to change the sound, put your own MP3 into the www subfolder and update the ALARM_FILE value. Upload the sketch to your Arduino Yún via Wi-Fi. If necessary, update the time zone in the web admin panel accessible at http://arduino.local.

6 To use the S.M.A.R.T. Alarm Clock, add an event to your Google Calendar for early the next morning. By default the alarm clock will only look at meetings before noon and set itself to go off one hour before the earliest meeting. You can change this behavior by adjusting the #define values ALARM_LATEST_HOUR and ALARM_BUFFER_MINS in the sketch. If the alarm goes off, press anywhere on the touchscreen to stop it. To test the email functionality, try sending yourself an email with the text "WAKE UP" in the subject. Now your S.M.A.R.T. Alarm Clock is fully configured and ready to wake you whenever it's needed—and let you get some sleep when it's not.

radioshack

CONTENTS

COLUMNS

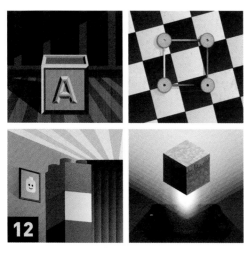

12

32

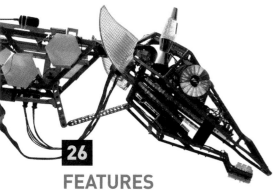

26

FEATURES

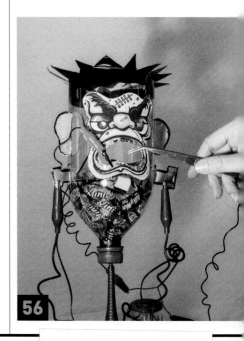

56

FIRST Robotics Teams use **ShopBots** to Help Them Become Champions!

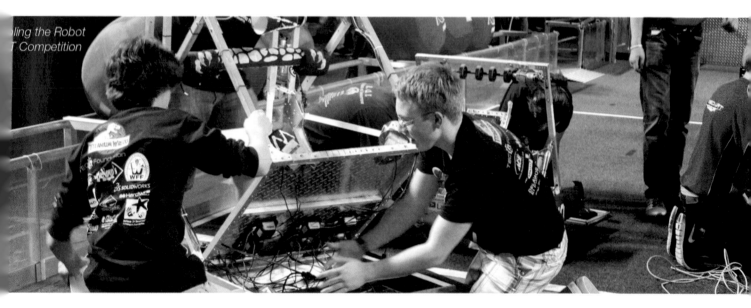

ling the Robot
T Competition

Chris Thompson, a self-described lifelong maker, trained as an engineer at the Naval Academy, saw 7 years of active duty and is now a Navy civilian engineer. Feeling the need to empower more young people with STEM skills that will help them succeed, Chris started the non-profit TEC Foundation, whose mission is to give everyone access to the tools of modern innovation.

Who: TEC Hub
Site: www.tecfound.org
Where: Eastern Shore of MD & VA
Tool: Full-size PRSalpha CNC, Handibot® Smart Tool

With community support, they secured funds for four full-size ShopBot CNC tools. As Chris puts it, "These are the cornerstone tools. They've transformed our FIRST Robotics teams' ability to really get engaged, use CAD, work out their robot designs and construct them."

> "I saw a need for an entity that fills a gap in our rural schools' education. TEC Hubs are places where you can learn and use any manner of technology."

TEC Hub's FIRST Robotics Teams have already medalled at Regionals and even gone on to National Competition. They've brought the portable Handibot® Smart Tool to competition, where students offered pit-crew assistance to teams needing emergency parts CNC'd.

TEC Hub student cuts aluminum robot parts on full size ShopBot

Chris believes the Handibot Tool can be a game changer in education. "The affordability of the tool is going to help a lot of less affluent communities gain access to CNC. Now any kid from anywhere can get excited about designing and engineering."

TEC Hub mentors make the difference!

earn more about Chris, Paul Suplee, Will Mast, Dave Miles
nd Greg Armstrong, whose combined NASA, Navy, and
ulinary experience has motivated kids to learn and make:
00kschools.com/blog   Search: TEC Hub

We make the tools for making the **future.**

ShopBot ®

888-680-4466 • ShopBotTools.com

SKILL BUILDERS

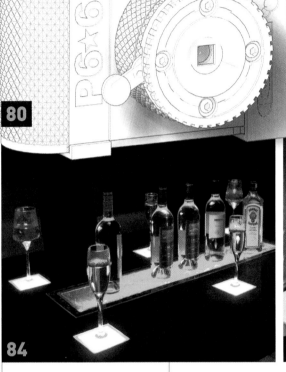

80

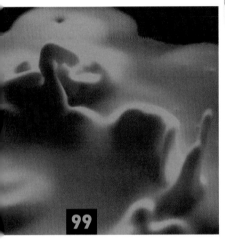

99

PROJECTS

84

90

TOOLBOX

OVER THE TOP

NEAT!

95

Vol. 41, September 2014. *Make:* (ISSN 1556-2336) is published bimonthly by Maker Media, Inc. in the months of January, March, May, July, September, and November. Maker Media is located at 1005 Gravenstein Hwy. North, Sebastopol, CA 95472, (707) 827-7000. SUBSCRIPTIONS: Send all subscription requests to *Make:*, P.O. Box 17046, North Hollywood, CA 91615-9588 or subscribe online at makezine.com/offer or via phone at (866) 289-8847 (U.S. and Canada); all other countries call (818) 487-2037. Subscriptions are available for $34.95 for 1 year (6 issues) in the United States; in Canada: $39.95 USD; all other countries: $49.95 USD. Periodicals Postage Paid at Sebastopol, CA, and at additional mailing offices. POSTMASTER: Send address changes to *Make:*, P.O. Box 17046, North Hollywood, CA 91615-9588. Canada Post Publications Mail Agreement Number 41129568. CANADA POSTMASTER: Send address changes to: Maker Media, PO Box 456, Niagara Falls, ON L2E 6V2

PRESENTED BY Disney

WORLD
NEW YORK
HALL OF SCIENCE
Maker Faire®
YEAR OF THE MAKER

CELEBRATING COMMUNITY AROUND THE WORLD

nysci

BAY AREA NEW YORK

GREATEST SHOW & TELL ON EARTH

5th Annual NEW YORK SEPT 20+21
makerfaire.com Brought to you by MAKE magazine

Make:

FOUNDER & CEO
Dale Dougherty
dale@makezine.com

CFO
Todd Sotkiewicz
todd@makezine.com

CREATIVE DIRECTOR
Jason Babler
jbabler@makezine.com

VICE PRESIDENT
Sherry Huss
sherry@makezine.com

EDITORIAL

EXECUTIVE EDITOR
Mike Senese
msenese@makezine.com

COMMUNITY EDITOR
Caleb Kraft
caleb@makermedia.com

MANAGING EDITOR
Cindy Lum

PROJECTS EDITOR
Keith Hammond
khammond@makezine.com

TECHNICAL EDITOR
David Scheltema

DIGITAL FABRICATION EDITOR
Anna Kaziunas France

EDITORS
Laura Cochrane
Nathan Hurst

EDITORIAL ASSISTANT
Craig Couden

COPY EDITOR
Laurie Barton

PUBLISHER, BOOKS
Brian Jepson

EDITOR, BOOKS
Patrick DiJusto

DESIGN, PHOTOGRAPHY & VIDEO

ART DIRECTOR
Juliann Brown

DESIGNER
Jim Burke

PHOTO EDITOR
Jeffrey Braverman

PHOTOGRAPHER
Gunther Kirsch

MULTIMEDIA PRODUCER
Emmanuel Mota

VIDEOGRAPHER
Nat Wilson-Heckathorn

FABRICATOR
Daniel Spangler

WEBSITE

MANAGING DIRECTOR
Alice Hill

DIRECTOR OF ONLINE OPERATIONS
Clair Whitmer

WEB PRODUCERS
Bill Olson
David Beauchamp

SALES & ADVERTISING

SENIOR SALES MANAGER
Katie D. Kunde
katie@makezine.com

SALES MANAGERS
Cecily Benzon
cbenzon@makezine.com
Brigitte Kunde
brigitte@makezine.com

CLIENT SERVICES MANAGERS
Mara Lincoln
Miranda Mota

MARKETING COORDINATOR
Karlee Vincent

COMMERCE

VICE PRESIDENT OF COMMERCE
Kelly Peters

DIRECTOR OF SHED DESIGN
Riley Wilkinson

RETAIL CHANNEL DIRECTOR
Kirk Matsuo

PRODUCT INNOVATION MANAGER
Michael Castor

MARKETING

VICE PRESIDENT OF MARKETING
Vickie Welch
vwelch@makezine.com

MARKETING PROGRAMS MANAGER
Suzanne Huston

MARKETING SERVICES COORDINATOR
Johanna Nuding

MARKETING RELATIONS COORDINATOR
Sarah Slagle

DIRECTOR, RETAIL MARKETING & OPERATIONS
Heather Harmon Cochran
heatherh@makezine.com

MAKER FAIRE

PRODUCER
Louise Glasgow

PROGRAM DIRECTOR
Sabrina Merlo

MARKETING & PR
Bridgette Vanderlaan

CUSTOM PROGRAMS

DIRECTOR
Michelle Hlubinka

CUSTOMER SERVICE

CUSTOMER SERVICE REPRESENTATIVE
Kelly Thornton
cs@readerservices.
makezine.com

Manage your account online, including change of address:
makezine.com/account
866-289-8847 toll-free in U.S. and Canada
818-487-2037,
5 a.m.–5 p.m., PST
makezine.com

PUBLISHED BY

MAKER MEDIA, INC.
Dale Dougherty, CEO

Copyright © 2014
Maker Media, Inc.
All rights reserved.
Reproduction without permission is prohibited.
Printed in the USA by Schumann Printers, Inc.

CONTRIBUTING EDITORS
William Gurstelle, Nick Normal, Charles Platt, Matt Richardson

CONTRIBUTING WRITERS
Johannes Baiter, David Bakker, Noah Bicknell, Sebastien Bridonneau, Glen Bull, Kathy Ceceri, Eric Chu, Jean Consorti, Stuart Deutsch, Katy Franco, Paul Gentile, Gretchen Giles, Hossein Haj-Hariri, Jim Hannon, Gregory Hayes, Jessica Hendricks, Matti Kariluoma, Bob Knetzger, Lindsay Lawlor, Wayne Losey, Bruce Lund, Anne Mayoral, Forrest M. Mims III, Goli Mohammadi, Riley Mullen, Abrianna Nelson, Sean Michael Ragan, Daniel Reetz, Conor Russomanno, Dan Shapiro, Todd Schlemmer, Lee Siegel, Debbie Sterling, Matt Stultz, Pete Sveen, Alice Taylor, Elliot Williams

CONTRIBUTING ARTISTS
Matthew Billington, Nate Van Dyke, Bob Knetzger, Rob Nance, Damien Scogin, Julie West

ONLINE CONTRIBUTORS
Alasdair Allan, Jimmy DiResta, Agnes Niewiadomski, Haley Pierson-Cox, Andrew Salomone, Andrew Terranova

ENGINEERING INTERNS
Enrique DePola, Paloma Fautley, Sam Freeman, Andrew Katz (jr.), Pierre-Alexandre Luyt, Brian Melani, Nick Parks, Sandra Rodriguez, Sam Scheiner

Comments may be sent to:
editor@makezine.com

Visit us online:
makezine.com

Follow us on Twitter:
@make @makerfaire
@craft @makershed

On Google+:
google.com/+make

On Facebook:
makemagazine

CONTRIBUTORS

What is your favorite tool and why?

Matthew Billington
Toronto, Canada (Tinkering Toys illustration)

My favourite tool would be my HP scanner from 2001. I drag it around all over the country, expose it to airport security, force it to scan as many images as it can muster. It sounds like an old jalopy. Sure I could buy another, but it wouldn't feel like MY scanner!

Glen Bull
Charlottesville, Virginia (The Lab in the Classroom)

Fab@School Designer, a CAD program we are developing for the Laboratory School for Advanced Manufacturing. It encourages rapid prototyping through multiple iterations using 2D fabricators like computer-controlled die cutters. The working design is then manufactured in its final form with a 3D fabricator.

Ann Mayoral
San Francisco, California (What is a "Girl Toy"? and Beyond Barbie)

My 30-piece ratcheting screwdriver set. It used to belong to my husband, but I borrowed it so often during the build of our 3D printer that it just never got back to him. Its small size and versatility make it my go-to tool. I heart Torx!

Pete Sveen
Bozeman, Montana (LED Concrete Patio Table)

My Hobart 250ci plasma cutter. It's super simple to use, and I can make metal art and fabricate metal into just about anything I can dream of. This tool makes cutting metal easy and it's a ton of fun to use!

Andrew Salomone
Brooklyn, New York (How to Print a One-Page Book by Hand)

My favorite tool is a utility knife because I almost never make a project that doesn't require one.

PRINTED WITH
SOY INK

Why the Banana Piano Makes Sense

BY DALE DOUGHERTY, founder and CEO of Maker Media.

Jay Silver

BLINDFOLDED, THE CHILD SAT AT A TABLE BEFORE A SET OF TOY BRICKS AND ANOTHER SET OF CUBES. The child was asked to pick them up and sort them, putting bricks on one side and cubes on the other. It's a simple but useful task for a 3-year-old, one of many object-based exercises that Maria Montessori, the 20th-century Italian physician and educator, used to develop an "education of the senses" that would help children integrate their own experience of the world. Montessori noted that children "are very proud of seeing without their eyes, holding out their hands and crying 'I can see with my hands.'"

Montessori describes other exercises that encourage children to explore the sense of touch: setting out metal containers of water heated at six degree intervals; tablets made of three different woods that differ in weight by six grams; other tablets that have alternating strips of smooth paper and sandpaper. Children were asked to recognize the differences and place the objects in some order. It is rudimentary hands-on learning to engage the senses.

It reminds me of a felting activity I saw at a Mini Maker Faire. Children would dip sheep's wool into a pan of soapy water and then pull it out to shape it. As I watched, they wanted to play in the sudsy waters, splashing, creating bubbles, and causing the water to rock from one side to the next. The experience was messy and fun, and I wondered if this kind of play was new to them. Did they not do this at home?

I have said, somewhat jokingly, that young kids today seem to have "tactile deficit syndrome," and I get nods from people. Today's toddlers are growing up with the touchscreen interface and its look and feel. While a touchscreen responds to touch, it provides almost no tactile feedback. The iPad is more like a remote control, and children using it sense the world with eyes and ears alone, much like TV.

Montessori believed that this education of the senses was important for the child's ongoing development. "To teach a child whose senses have been educated is quite a different thing," she wrote. "Any object presented, any idea given, any invitation to observe, is greeted with interest, because the child is already sensitive to such tiny differences as those which occur between the forms of leaves, the colours of flowers, or the bodies of insects." One of the hallmarks of the method is to structure the environment and teaching so that children become independent, self-directed learners. It leads to agency, not apathy.

This philosophy of education later became known as "constructivist." Learning is active — we don't receive knowledge, we must construct the world to understand it, as with building blocks. In the words of Swiss psychologist Jean Piaget, "to invent is to learn." At MIT in the 1980s, Seymour Papert, who believed that computer technology was a tool for children to begin teaching themselves, developed the idea of "constructionism." Papert argued that children construct knowledge best when they are constructing something real. One of Papert's students was Mitch Resnick, who developed the Lifelong Kindergarten program at MIT Media Lab to explore the role of technology in learning. Resnick developed the Computer Clubhouse in the 1990s and, more recently, the Scratch programming environment for children.

In a paper, Resnick asks the Sesame Street question: Which of these things is not like the other — computer, television, or finger painting? He believes the nonobvious answer is television. "Until we start to think of computers more like finger paint and less like television," he writes, "computers will not live up to their full potential." Neither will our children.

Jay Silver, a student of Resnick's at MIT, followed the strands of constructivism and constructionism while at the Media Lab. Silver's unfinished doctoral thesis is titled "World as Construction Kit."

While working on his thesis, his Makey Makey project took off — big time. Silver and co-inventor Eric Rosenbaum raised more than $500,000 on Kickstarter in 2012, and Makey Makey began spreading as an entry-level system for young makers. At a recent teacher's conference, I saw educators demonstrating how to use it in the classroom. In June, Silver was invited to bring Makey Makey to the White House Maker Faire. He showed up in his usual colorful T-shirt and baggy shorts.

Makey Makey changes the interface for computing to almost anything you want, and it is a game changer. Children and adults can interact with computers in new, creative ways. At MakerCon, where Silver spoke, a computer snapped our photo at the moment that we high-fived each other. While one hand touched a wire, our other hands completed a circuit when they slapped, tripping the shutter. With Makey Makey, you don't have to use a keyboard. You can slap hands. To play a piano on your computer, you can tap the skin of a banana. You can even do it blindfolded.

That you can create a banana piano with Makey Makey is a seemingly silly thing — but it is also surprisingly important. Silver calls it an "invention kit," a new kind of toy or game for "the simultaneous combination of exploration and creative action that leads to a new way of seeing the world." The banana piano opens up unexplored possibilities for interactions between computers and humans that have a touch and feel. It demonstrates that "computers can also be used as a 'material' for making things," as Resnick wrote. The material world can be organized to interact creatively with computers. It's what the touchscreen-fixated generation needs — preparing them to see with their mind's eye. ●

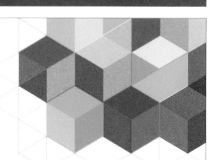

Young Makers and Safety Fundamentals

» On June 18, 2014, *Make:* magazine celebrated the National Day of Making (see our White House Maker Faire story on page 26) with a Meetup at the *Make:* office. One guest was 11-year-old Riley Mullen, who drove his Roomba-hacked robot right into the *Make:* Labs. Riley also brought his Mini Spin Art project, featured on page 95. Riley sent all of the editors handmade cards (example of artwork and message at left and below) featuring his artwork after the event. Thank you, Riley, for inspiring us with your creativity and ingenuity.

Dear David (Scheltema — Technical Editor),
Thank you for the behind-the-scenes tour of Make's office. I really enjoyed seeing the laser cutter and magazine layout. Your time and kindness were greatly appreciated. Thank you also for letting me scavenge through the "junk" bin. It was like Christmas!

Sincerely, **Riley Mullen**

» Hi! I love your magazine! I bought issue 1 as soon as I discovered it and have never looked back. I noticed on page 102 in Volume 40, you feature a close-up of a gloved hand using the table saw. Wearing gloves while using table saws is very, very dangerous. Gloves aren't nearly tough enough to prevent the table saws from cutting you, and they can easily get snagged on the blade and drag your whole hand in.

— *A Concerned Reader*

» As a journeyman tool and die maker (who also teaches engineering at Ohio State) I'm a bit concerned about some of the photos in Volume 40. On the cover, and inside the magazine, you show photos of Emily Pilloton wearing a ring made from a hex nut. Any machinist will tell you that wearing rings, especially large clunky rings, around rotating machinery is a damned good way to lose a finger.

Likewise, on page 57 you show a young woman with long hair running a band saw. A female grad student was killed when her ponytail became tangled in a metal lathe in a tragic accident at Yale a few years ago. In our shops we insist that students with long hair pull it back and tie it up.

Having worked in a tool room at GM for several years, I can tell you industrial accidents are generally life-changing events. We require all our second-year mechanical engineering students to take a "maker course," but we're absolute fanatics about safety. Allowing a young person to be permanently maimed by a machine tool is not something I would want on my conscience.

— *Blaine Lilly, Columbus, Ohio*

MAKE: LABS MANAGER SAM FREEMAN RESPONDS:

» Thank you both for providing great feedback and valid safety points (the Yale incident shook the entire community). We take safety seriously in our own *Make:* Labs, and everyone working in it is instructed on our safety procedures, which include (among other things) the following:

• Never wear gloves around spinning parts. This goes for saws, drills and grinders.
• Likewise, long sleeves should be avoided or rolled up, and no loose jewelry or cords. Wrist watches, wedding rings, and hair must all be securely out of the way.
• Gloves should be worn when welding and for high-temperature operations, and also when handling rough materials.
• When using power tools, have someone around who can get help in an emergency.
• Safety glasses should be worn any time there's a projectile risk. This includes all power tools, cutters, and clips.
• Hearing protection should be worn any time there's significant noise. OSHA recommends protection for anything above 89dB, and rates a hand drill at 98dB.
• No open-toe footwear.

Other great safety manuals include:
web.stanford.edu/dept/EHS/prod/aboutus/documents/safetyman/toolsafety.html, and osha.gov/Publications/OSHA3170/3170-02R-2007-English.html#Controlling9

» In receiving my new Aug/Sept issue of *Make:* magazine today, I immediately went to the Emily Pilloton article on hands-on learning. While I laud her initiative and efforts with her program, I was a little disappointed. Perhaps I'm just having an off day? I have been working for five years now in urban public school teaching Project Lead the Way curriculum that I feel is similar, but vastly broader in scope. My kids design and build a wide variety of hands-on projects utilizing multiple technical disciplines. I'd encourage *Make:* to take a look at PLTW in the future. It's worthy!

— *Michael David Wheeler*

EXECUTIVE EDITOR MIKE SENESE RESPONDS:
» Thanks for the note — we're obviously huge fans of what Emily and Project H are doing, and we're also always interested in hearing about what other organizations and projects are underway to help inspire and equip the next generation to be hands-on with tools and technology. Drop us a line at editor@makezine.com.

» Dear Editor(s), I liked your article "Grant Imahara's Hollywood Dream Machine." I liked reading about the different types of machines he helped on or made. I also liked this because I am on an FTC (FIRST Tech Challenge) Team. It is fun to see that the skills that I'm learning can be used in real life. —*Sincerely, Jared N.*

MAKE AMENDS:

■ HackPittsburgh pointed out that in Vol. 40's Most Interesting Makerspaces, we erroneously list HackPittsburgh in Pittsburgh, Philadelphia, rather than Pittsburgh, Pennsylvania. All *Make:* Edit staff report for Remedial Geography class.

Experience Over Entertainment

How the toy industry is growing in the right direction. Written by Wayne Losey ■ Illustration by Jim Burke

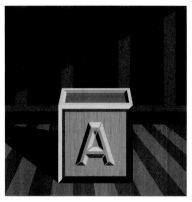

THE NATURE OF PLAY IS EVOLVING IN OVERDRIVE. For years, kids have been migrating away from the historical toy categories that made the industry an economic juggernaut. With the rise of mobile, interactive screenplay has pulled a lot of users away from analog toys. With their low cost, high production values, deep interactivity, and (of course) mobility, they are, literally, the killer app for play. Compressing all that play value into one experience has made it difficult for the toy industry — and analog toys — to keep up.

But the reality is more complex. The major shift isn't economic; what has changed is user expectations. Continued exposure to the digital world has rewired our brains to ask and require more of our products. That's not necessarily a bad thing.

We don't have to look far for examples of how quickly things have changed. Watch a young child swipe any common TV or laptop screen. After a few quick attempts, you'll see a mixture of frustration, confusion, and contempt. The interaction exposes that widening divide between the deep, dynamic play of the digital world and the mostly static world of consumer products.

Toys, like many traditional products, are being disrupted. But mobile apps aren't killing the toy business; they're forcing it to evolve. What is working well in the toy business are both classic and next-gen products that offer many of the same values that are effective in mobile — self-expression, social engagement, intellectual complexity, challenge, mastery, and a wide degree of freedom. Both digital and physical toys are learning that experience is more important than entertainment.

Playable, creative platforms are thriving in what is still a tough economy. The sales of Crayola, Lego, Imaginext, Play-Doh and Magic: The Gathering are booming because they leave room for the child's desires without limiting them to a single, close-ended reality. These platforms function more like tools than traditional products: Like any great tool, their value is derived from what you can make with them.

Another seismic shift in the toy economy is in diversity. We've seen a boom in indie and boutique toys over the last five years, giving purchasers more choice and resetting the predictable pricing and product features built upon international-scaled business practices. The impact is very similar to what happened in the comic book industry in the '80s, when independent creators brought a lot of variety and innovation into an industry that was stalled and mostly dominated by superheroes since the '60s.

Kickstarter and other crowdfunding platforms have made it possible to cultivate and sell directly to niche tastes. We've seen a huge increase in modular action figures, construction toys, and an amazing spectrum of miniature and board games, almost none of which would have been possible prior. Every successful launch redefines norms and forces industry leaders to focus on a new path forward.

What's most exciting to see is the coming impact of maker-driven development tools and technology. The variety of items you can create with just a small arsenal of equipment — such as a 3D printer, Shapeoko, and TinyDuino — is nearly limitless. For the first time in history, prototyping isn't a huge barrier to entry.

The emergence of these tools and consumer's growing thirst for new approaches is creating a virtuous cycle of escalating innovation. We're all part of a growing ecosystem, designers and makers using new tool sets to create next-gen tools for budding designers, who will create even better tools. The network effect of the web blurs the lines between professional and amateur, building new communities and driving markets to become richer and more diverse. Users will buy fewer premade products and instead spend money and time creating their own awesome playthings. Consumers will become producers and toolmakers. And that is a future of toys we can support. ❷

WAYNE LOSEY is a veteran toymaker and Chief Creative for Modio, a new app that enables users to design, customize, and 3D print their own unique, poseable characters.

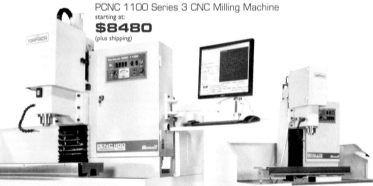

What Is a "Girl Toy"?

Written by Anne Mayoral

The toys we buy (and make) are sending a message. Make sure it's the right one.

TAKE A 15-MINUTE TRIP TO A BIG-BOX STORE AND YOU'LL SEE THE GIRLS' AISLE DOMINATED BY PINK TOYS — dolls, play sets, games, and dress-up clothing, all in various shades of pink. Aside from the color, the toys share another similarity: They all encourage specific patterns of play that focus on role-play, nurturing, and domestic crafting. Conversely, the boys' aisle is full of action figures, erector sets, blasters, and cars — toys that highlight action, building, and violence.

Gender-based toy marketing is nothing new, but the impact on our children shapes their futures. According to the National Institute for Play, "play shapes our brains, creates our competencies, and ballasts our emotions." If we limit the play patterns for girls through the use of segregated toys, they will experience only a narrow view of the world, suggesting each gender can excel only in certain areas.

For girls, the message is: I cannot be good in math and science. In a culture charged with strong gender expectations and social pressures, young girls develop a distorted idea of their expected role in life and strive to live up to skewed ideals that fall short of their potential. The message has been so pervasive that today, women are heavily under-represented among working engineers, scientists, and mathematicians.

There are a few new engineering toys specifically designed for girls that are aimed at combatting this effect. GoldieBlox and Roominate strive to capture girls' imagination before they lose interest in science and math. Created by female engineers, these toys focus on building, designing, and innovating, play patterns usually offered only to boys. However, they

ANNE MAYORAL is a rocket scientist turned industrial designer who enjoys demystifying science and math. She currently works as a freelance designer and teaches hands-on STEAM classes to girls in order to cultivate a love of science. See her flowchart for finding good toys for girls on page 54.

also incorporate elements of traditional girl toys, like dollhouses, parades, and tree houses. Some critics worry that the pastel colors and overall premise — girls can only tackle girly problems — further highlights the gender messaging that plagues the rest of the toy aisle. Still, parents often see these toys as a refreshing gateway to more advanced interests, teaching skills and concepts that might lead to STEAM fields.

In addition to engineering toys targeted at girls, some toymakers are creating gender-neutral toys. If we want to transform the toy market, we need to break the rules and work to eliminate gender-typed toys, and support toymakers who do the same.

Still, gender-specific toys are pervasive, and it is difficult to find mainstream toys that are designed for both sexes and endorsed by both parent and child. As we design, make, and buy STEAM (science, technology, engineering, art, math) toys, we should think about the message they send to children.

Toys and activities that support types of play that defy gender stereotypes will teach the skills, experiences, and intuition that foster an aptitude for STEAM fields. When faced with challenges in math and science, girls will receive a message of competence and confidence.

There's no easy solution; it'll take time, and change will have to come one girl at a time. We can make a more immediate and significant impact with our girls by encouraging them to play with whatever toys interest them, despite their color.

The key for parents is not to push their daughters one way or the other — not every girl should be, or will want to be, an engineer. Instead, the goal is to offer a variety of toys and a chance for hands-on discovery and making, either at home or in a local makerspace, and then let them choose for themselves. If we foster these skills, it will give them the chance to explore their own interests, not limited to the opportunities defined by their gender. ◐

Jeffrey Braverman

MAKER FAIRES
around the world

SN Jacobson

COMMUNITY-BASED, INDEPENDENTLY PRODUCED EVENTS ARE HAPPENING ALL OVER THE GLOBE. JOIN ONE!

MAKER FAIRE ORLANDO
September 13-14, Orlando, FL
Held at the Orlando Science Center, this year's faire will be bigger than ever. Featuring Power Racing Series, Nerdy Derby, robots, art, and more.
makerfaireorlando.com

MAKER FAIRE SILVER SPRING
September 14, Silver Spring, MD
Presented by KID Museum, this year's faire aims to surpass last year's with robots, upcycled toys, 3Dprinters, a giant marble run, and more. Plus, it's free!
makerfairesilverspring.com

✪ ✪ WORLD MAKER FAIRE NEW YORK ✪ ✪
September 20-21, Queens, N.Y.
Our East Coast flagship Faire is now in its fifth year and still growing, boasting 60,000 attendees and 700 makers in 2013. Held at the New York Hall of Science, this is the premier showcase of East Coast maker innovation and creativity.
makerfaire.com

MAKER FAIRE MILWAUKEE
September 27-28, West Allis, WI
Hosted by the Betty Brinn Children's Museum and the Milwaukee Makerspace, the two-day event will feature more than 100 makers at the Wisconsin State Fair Park.
makerfairemilwaukee.com

MAKER FAIRE ROME
October 3-5, Rome, Italy
Last year's faire hosted 200+ makers and 30,000 attendees — and this year has grown to include forums about open hardware, maker cities, and more.
makerfairerome.eu

MAKER FAIRE ATLANTA
October 4-5, Decatur, GA
After last year's one-day event drew 125 makers and brought in 10,000 people from across the southeastern U.S., this year they are moving to a shiny new location and adding a second day to the event. Best of all, attendance is free!
makerfaireatl.com

MINI MAKER FAIRE CALENDAR MID-SEPTEMBER THROUGH NOVEMBER 2014

• **SEPTEMBER 13**
Greenbrae Mini Maker Faire (CA)

Nashville Mini Maker Faire (TN)

Omaha Mini Maker Faire (NE)

Waterloo Mini Maker Faire (Ontario, Canada)

• **SEPTEMBER 13-14**
Albuquerque Mini Maker Faire (NM)

Cincinnati Mini Maker Faire (OH)

Fort Wayne Regional Maker Faire (IN)

Portland Mini Maker Faire (OR)

• **SEPTEMBER 19-21**
Lisbon Mini Maker Faire (Portugal)

• **SEPTEMBER 20**
Prince George Mini Maker Faire (Canada)

• **SEPTEMBER 20-21**
Kerkrade Mini Maker Faire (Netherlands)

Maker Faire Seoul (Korea)

• **SEPTEMBER 26**
Tartu Mini Maker Faire (Estonia)

• **SEPTEMBER 27**
León Mini Maker Faire (Spain)

Louisville Mini Maker Faire (KY)

Salt Lake City Mini Maker Faire (UT)

• **OCTOBER 4**
Charlottesville Mini Maker Faire (VA)

Greater Portland Mini Maker Faire (ME)

Inland Empire Mini Maker Faire (Riverside, CA)

MIT Mini Maker Faire (Cambridge, MA)

NoCo Mini Maker Faire (Loveland, CO)

Scranton Mini Maker Faire (PA)

• **OCTOBER 4-5**
Champlain Mini Maker Faire (VT)

• **OCTOBER 11**
Chattanooga Mini Maker Faire (TN)

Columbus Mini Maker Faire (OH)

• **OCTOBER 18**
Akron Mini Maker Faire (OH)

Colorado Springs Mini Maker Faire (CO)

• **OCTOBER 18-19**
Hong Kong Mini Maker Faire (China)

• **OCTOBER 19**
East Bay Mini Maker Faire (CA)

• **NOVEMBER 1**
Houston Mini Maker Faire (TX)

• **NOVEMBER 8**
Miami Mini Maker Faire (FL)

MADE ON EARTH

The world of backyard technology

Know a project that would be perfect for Made on Earth?
Email us: *editor@makezine.com*

LOS FERRONAUTAS

SEFT1.NET

Mexico is crisscrossed by a vast network of abandoned railroads, no longer connecting many communities that had once relied on their freight and passenger service. In 2006, **Ivan Puig** and **Andrés Padilla Domene**, brothers and artists in Mexico City, cooked up the idea to explore the lost routes and places.

"We designed a spacecraft, a research tool that would take us on an expedition to the inner space of our country," says Puig. Domene adds, "Its design revisits the idea of the future as seen from the past." Built atop the chassis of a convertible railway pickup truck — able to run on rails or land — the aluminum fuselage houses a handsome combination of analog and digital technology, plus a cozy living space for the two "ferronautas," or railnauts.

When their team finished the construction of the SEFT-1 (Sonda de Exploración Ferroviaria Tripulada) in 2010, the brothers embarked on a trip through Mexico, guided by modern and 19th century maps. Between eating and sleeping in the SEFT-1, they recorded remnants of the old railroad — from brightly repurposed stations and routes to dark ghosts of former prosperity, sometimes mere memories with no physical trace at all. As they rolled along, they collected and shared their tales, images, route information and more. A stunning collection from this and other journeys is now available on their interactive website.
—*Gregory Hayes*

SEFT-1 project: Ivan Puig and Andrés Padilla Domene

TREETOP TINKERING

Perched high in a coniferous tree in Sandpoint, Idaho sits a hexagonal treehouse, handmade by 23-year-old **Ethan Schlussler**. During the build process, he grew tired of climbing up and down the ladder over and over So he created a treehouse elevator, using his mom's old bicycle.

The elevator is a pulley system that uses a water heater as the counter balance. He modified the bike by moving the large sprocket from the front of the bike to the back to get a low enough gear. It takes him about 15 seconds to pedal the 30 feet up to his small front porch.

The treehouse is made with lumber Schlussler hand-milled himself and preserved with Danish oil. He invented the friction system that keeps the treehouse aloft, which he describes as "experimental." The roof is self-supported, which means it can move independently of the walls, to flex in the wind and twist with the tree.

Two nearby trees with platforms connect to the treehouse via cable bridges, and there is a zip line that leads directly across the forested yard to his mom's house. And the clever ideas just keep coming: Next, Schlussler plans to build a pedal-powered trolley car that he can ride back up the zip line.

—*Laura Cochrane*

Steven Scarcello

Sean Fannin

NIXTROLA ELECTROTHERAPYDESIGNS.COM

"I went online to buy a nixie clock and didn't see any I really liked," says Woodland Hills, California tinkerer **Sean Fannin** about timepieces made from coldwar-era numeral-displaying nixie tubes. "A lot were DIY kits that ended up looking cheap. I wanted to build something a little more slick."

After buying and assembling a basic kit clock, Fannin envisioned a housing modeled after a Victorian phonograph. "I struggled trying to fit the nixie tubes in. Sitting at my desk, trying to draw it out — it was not working." He let the idea rest and waited for inspiration. Then one day, "I was watching

Cosmos episode 6. This CGI sequence about machine-generated chloroplast had brass bells with copper tubing, and it dawned on me. I sketched everything out in about five minutes."

After some desoldering and tinkering (about 40 hours worth, estimates Fannin, much of that learning curve), the resulting contraption houses the clock's components in a handsome new arrangement he calls the Curious Nixie. That original has spawned several evolutions, which Fannin sells on Etsy along with other works.

—Gregory Hayes

MAGIC CARPET JOEANDNATHAN.COM

When **Jonathan Bréchignac** isn't busy creating digital works for clients of his Paris design studio, he's on an analog meditation of pen and paper, elevating the humble Bic ballpoint to a next-level art medium with his incredibly detailed and meticulously drawn series titled *The Carpets*. Intended to approximate the size of Muslim prayer rugs, the smallest in the series, *Carpet n°3*, is roughly 37"× 23", and the largest, *Carpet n°1*, is 46"× 29".

Each carpet is developed and drawn organically, bit by bit, as a nod to ancient artisans who would spend years working on one piece of art. Bréchignac's first carpet took 15 months to draw, while subsequent carpets have taken roughly 6–8 months apiece. When asked if he employs any digital tools, he replies, "No, I draw everything directly on the paper. I just need a compass and a ruler."

To add what he calls "a 2.0 dimension" to the drawings, Bréchignac has penned in QR codes that correspond to pages on thecarpets.net, offering a "digital and evolutionary extension of the drawings."

—*Goli Mohammadi*

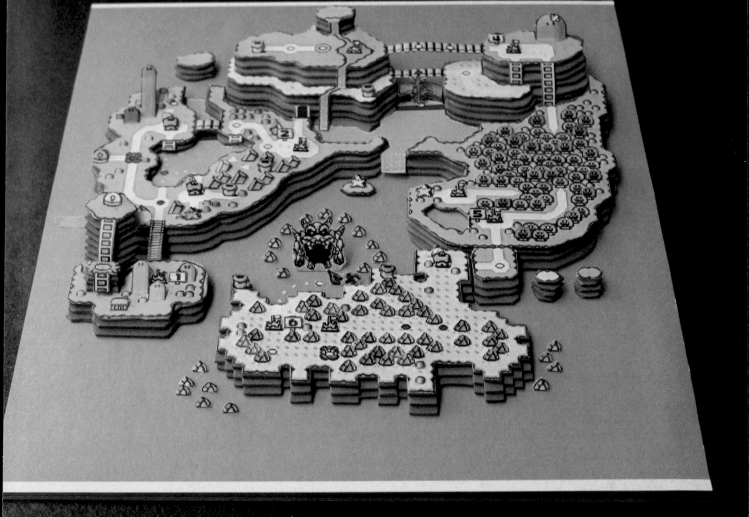

Alain Wildgen

CLASSIC VIDEO GAMES LEVEL UP

WUPPES.TUMBLR.COM

Classic video games are often viewed through the rosy colored glasses of nostalgia, pining for those lazy days of our youth when we could explore new universes for hours on end without a worry in the world.

Luxembourg-based **Alain Wildgen** pulled off his rose-colored glasses to pay a very real and tangible homage to his favorite games. Constructing dioramas mainly out of foam and paper, he manages to add that extra bit of depth that makes these feel just right.

Wildgen touches up the original game artwork in Gimp, then prints everything out at the correct size. After that, he cuts pieces of foam by hand to get the exact heights and shapes he wants. When it all comes together, the result is simply stunning. Though he has had many requests for sales, Wildgen insists that these works of art are for his personal collection only.

—Caleb Kraft

THE LAB in the CLASSROOM

Written by Glen Bull, Hossein Haj-Hariri, and Abrianna Nelson

3D PRINTERS ARE COMING TO SCHOOLS. HOW DO WE MAKE THE MOST OF THEM?

Robert Munsey

DESPITE THE BILLIONS INVESTED IN EDUCATIONAL TECHNOLOGIES IN SCHOOLS EACH YEAR, few controlled research studies can document scalable gains in learning. New technologies do not automatically translate into improved learning. Desktop 3D printers have now become affordable enough to place in schools, which are acquiring them by the thousands, facilitated by crowd-sourced initiatives such as Donors Choose and Kickstarter.

In 2010, Hod Lipson, director of Cornell University's Creative Machines Lab, challenged a group of educators at the National Technology Leadership Summit to plan for the introduction of 3D printers to schools, asking: "If you could rewind the clock and introduce microcomputers in schools again, what might you do differently?"

What Lipson was asking is: How can emergent technologies be deployed to create new opportunities for effective and engaging learning in schools? As professors and faculty at the University of Virginia, we responded, collaborating with local schools to establish a K-12 Design Laboratory. It was to be a test bed for developing curricula based on digital fabrication, including 3D printing.

The promising results led to a joint venture between the Charlottesville and Albemarle school systems and the University of Virginia, established with support from grants from the National Science Foundation, the Commonwealth of Virginia, and local government. Two middle school sites, the Buford Engineering Design Academy and the Sutherland Engineering Design Academy, were launched at the beginning of the 2013-2014 school year.

A glimpse of a Lab School classroom reveals students working on projects in small groups, with 3D printers arrayed along the walls — one for every four students. Other fabrication technologies, such as computer-controlled die cutters, drawers of microelectronic components, sensors, motors and actuators, soldering stations, hand tools, and even laser cutters and CNC machines are readily available. Student projects are open-ended to incorporate engineering design into science teaching.

GLEN BULL is a professor of STEM Education in the Curry School of Education at the University of Virginia, and principal investigator for the Laboratory School for Advanced Manufacturing initiative.

HOSSEIN HAJ-HARIRI is a co-P.I. and chair of the University's Department of Mechanical and Aerospace Engineering.

ABRIANNA NELSON is the Lab School Liaison at the University of Virginia.

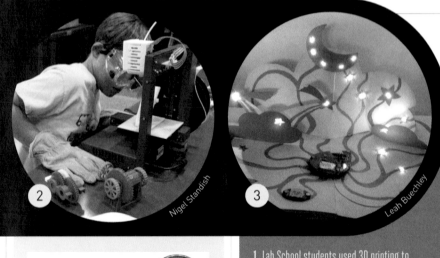

Nigel Standish

Leah Buechley

Courtesy of the Smithsonian's National Museum of American History

1. Lab School students used 3D printing to reconstruct the Morse-Vail telegraph relay.
2. A Lab School student examines a 3D print in progress.
3. An example of artwork created with circuit stickers.
4. Top: An 1885 reproduction. Bottom: Patent model of the 1854 Page motor, US Patent #10480.

It was here, with the help of curators from the Smithsonian's National Museum of American History, that two Lab School students, Jenn and Nate, reconstructed the Morse-Vail telegraph and relay. They used Alfred Vail's 1845 description of the device to design a modern version using digital fabrication technologies.

The work of makers such as Samuel Morse, Benjamin Franklin, Joseph Henry, and Alexander Graham Bell illuminates the process of invention and innovation. Their seminal discoveries are more accessible and easier for novices to understand than many modern techniques. The functions of the electromechanical systems of this era have tangible components that can be deconstructed and understood. During the initial years of the U.S. patent office, submission of a working model as a proof-of-concept was required for a patent. A selection of these patent models are now housed in the Smithsonian.

So the Smithsonian is collaborating with the Lab School to digitize key inventions for young makers like Jenn and Nate to reverse engineer. The Smithsonian's 3D explorer (http://3d.si.edu) allows students to measure every angle and even analyze cross sections of objects. As the Lab School inventions — from the telegraph to Charles Page's early electric motor and more — get digitized, the site will provide 3D files and supporting materials to help other schools replicate the process.

The goal is not an exact physical replica, but a reinterpretation of the device using modern manufacturing technology. The three-dimensional scans of the artifacts are inspiration for the students' own designs, allowing them to create a product that is uniquely their own.

The reconstruction helped Jenn and Nate understand the relationship between science and engineering. They applied the principles they learned in science class to the design of the telegraphic relay to determine how much current in the primary coil was needed to activate the secondary circuit.

They enjoyed the process of scientific exploration and discovery. They learned,

contrary to what they thought, that there are many things scientists do not know or understand. They saw that Vail and Morse experienced problems parallel to their own, both in science and engineering. In science: Neither the scientists (in 1840) nor the students (in 2014) fully understood the properties of electricity. In engineering: Both the inventors and the students had difficulty fabricating a reliable relay with a three-point connection.

A pilot cadre of students from the Laboratory School participated in an Engineering Design Academy this summer. The student engineers learned about telegraphic relays, solenoid engines, and linear motors. Teams were challenged to design and fabricate an electromechanical tone sequencer capable of playing a tune, and the students combined the relays, solenoids, and linear motors to create an electromechanical sequencer that reproduced the chimes of London's Big Ben, and presented their invention at the Smithsonian. In the future, exemplary inventions will be showcased in a Museum of Electronic and Moveable Objects in each school.

This work with electromechanical inventions serves as a springboard for inventions that incorporate modern technologies. or example, an interactive mural is being designed for each school that will incorporate circuit stickers — peel and stick electronics for crafting circuits — to recreate the students' electromechanical music machine.

The Lab School has already affected the educational directions of at least two students: Jenn has decided to become a biomedical engineer, and Nate has chosen to focus on mechanical engineering.

However, the point is not that all students should select careers in engineering; the goal is to ensure all students can explore desktop fabrication technologies. Like any other tools, they can be applied in myriad ways that enrich children's lives and make learning more engaging.

As the Lab Schools enter their second year, advanced manufacturing technologies are being incorporated throughout the physical science curriculum. Students are creating their own inventions, and sharing their work through FabNet, a network of schools collaborating to co-construct new ways of teaching and learning in this shared space. ◉

THE ELECTRIC GIRAFFE Goes to

Written by
Lindsay
Lawlor

WASHINGTON

When the President invites you to his house, you move

LINDSAY LAWLOR
currently lives out in horse country in Ramona, California, where the giraffe has a machine shop to live in. By day he keeps people from catching on fire, and by night, Lindsay can be found in his shop, listening to thumping dance music, TIG welding and machining new and wonderful things for kids both young and old.

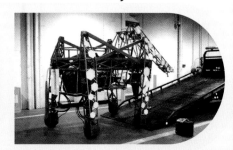

Public Lab

Ever since I unveiled him at Burning Man in 2005, Russell the Giraffe's been a crowd favorite — an interactive electric quadruped who speaks with a British accent. I've transported Russell, who is 18 feet tall and weighs one ton, to every Bay Area Maker Faire since the first, in 2006. But this May I received an unexpected message: Russell, his namesake programmer Russell Pinnington, and I were invited to the inaugural White House Maker Faire.

Obama has always been a fan of the self-employed entrepreneurs, the inventors, the makers. He chose to host a faire — in his house — and he wanted Russell.

You don't say no to the White House. But right away the distance and timeline were very daunting. Russell has never traveled farther than Nevada, never traveled alone, and never had such a tight schedule — less than 30 days to get from San Diego to Washington, D.C. I was filled with pride at having been invited, but could we pull this off in time?

Right away, Maker Faire's Sherry Huss and Louise Glasgow offered to help with funding, and I put together a plan for the trip to Washington. This was unknown territory; he'd be in a shipping container, all alone, with me nowhere near to keep an eye on him. As the momentum built, it became clear that there was no way I could handle things along with my day job as a fire alarm and security systems programmer. But my good friend Alan Murphy, who runs Murphy Surplus, said to not worry; he would front the money for the container and getting the giraffe shipped and handle all the logistics.

WITH NINE DAYS TO GO ...

Alan commissioned a custom ocean-going shipping container — a 10-foot version cut from a larger one — and had it shipped from Los Angeles. That gave us one evening to load it and only four days to make it to the Smithsonian. If anything went wrong at this point, the gig would be lost.

The container's inside was nine feet, six inches — tight, but with his neck and tail removed, Russell is exactly nine feet long. We slid the giraffe into the box, which still reeked of fresh paint from the previous day's construction, and he fit perfectly. The shipping company showed up early, and Russell began his four-day trip.

His human companions reached the Smithsonian's storage facility full of butterflies — in what condition would we find him? But the giraffe arrived exactly as he had been packed.

The crew at the Smithsonian had never seen anything like this. They milled about like excited kids as we extracted Russell from the container and put him back together. As I powered him up, checked his computer and hydraulic systems, and walked him back and forth, they stared in amazement at the robot with illuminated spots.

heaven and earth and an 18-foot robot to get there.

the electric giraffe project

1

2

3

CITIZEN MAKERS

From the very beginning, Maker Faire has been a wonderful celebration of makers and their projects. Taking that celebration to the White House in June recognized the incredible work of 60 or so notable makers. Each brought a project from home, got it through security, and set it up next to presidential portraits and busts. I was proud to be there and see the joy and gratitude on the faces of the makers, young and old. However, the White House Maker Faire also recognized the importance of what all makers are doing, and what they can do in America and the world. Citizen makers are introducing hands-on learning in education, creating new jobs in an innovation-driven economy, and generating new opportunities for civic engagement. The White House Maker Faire was a proud moment for the Maker Movement, and I couldn't help but think what an improbable journey it was for us all to be there.   — *Dale Dougherty*

What they were seeing was the culmination of years of work on a machine that walks on four tall legs, all made of hand-welded steel. Russell is radio-controlled and fully self-contained, running on 36 volts worth of deep-cycle batteries. His heart is a 3-horsepower electric DC motor that turns two hydrostatic drive pumps, which run at a constant speed and need only a simple servo to walk forward or backward. Everything else Russell does on his own, running Raffe-Ware, Pinnington's modular control program.

The day of the faire, we went to get the giraffe on a tow-truck rig. Right away we noticed a possible problem: The giraffe up on the rig was very high in the air — so high that we feared he might get stuck inside the loading dock and have to be taken apart again. But we rolled forward cautiously, and he cleared the door by a few inches.

That wasn't the only concern. Our driver spent a few minutes going over the route in his mind. There would be many overpasses, and a few long tunnels to go through. We estimated our height at around 13 feet, plus a few inches — approximately the maximum height of most diesel rigs.

Compound that with heavy traffic and little time, and we decided to risk it, with a top-heavy load, a driver who had presumably never driven a robotic giraffe, and a great deal of trepidation. The giraffe cleared the bridges with inches to spare, but the tunnels were perilous — boxed lights on the roof provided an obstacle course that he could crash into at any second. If he struck anything it would not only damage him for the show, but could rip him off the rig entirely.

Although Russell made it unscathed, we were not yet clear. We still had multiple layers of security to pass and no idea what the Secret Service would say about a huge robot giraffe heading for the White House.

In the inspection yard, the yardmaster approached me and asked for my paperwork. "I have none, sir," I shakily replied. After a minor interrogation and a review by the Secret Service, an armored SUV sped into the yard and out hopped another agent, with guns, body armor, and god knows what else strapped to his chest and belt. He said he would escort me to the White House immediately.

Here we were, a couple of guys who hacked together a pretty cool project, but ... the White House? Really? We were going in.

We set up the giraffe near the rose garden and waited. The heat was unbearable and most of the event moved indoors. And then he was there, the President of the United States. As he made his way up the lawn toward me, I just kept telling myself, "Don't pass out. Don't pass out."

"I like those ears!" he commented first, and I greeted him and shook his hand. I did my best to describe the giraffe, and I got him to pet it. "He has a bit of an accent," Obama noted, and we had a good chuckle about that. "Now, I hear you can ride this?" he asked, adding that the Secret Service would never allow it. He motioned me over for photos and put his arm around me.

I was arm in arm with the leader of the free world.

I can't thank everyone enough who helped make this happen. It was a dream that I never knew I had until the possibility arose. People ask me where we will end up next. "I don't know," I say. "I'm just holding onto his tail and he's dragging me along." ◗

1. Russell the Giraffe walks and/or rolls past the White House.

2. A guitar built by Thomas Singer as part of Sinclair Community College's STEM program in partnership with MIT's Mobile Fab Lab.

3. The first-ever 3D-printed bust of a U.S. president.

4. A touch-activated banana piano built with Makey Makey.

5 Russell in Washington.

OUT
OF YOUR
MIND

Rise of the Brain-Computer Interface

Written by Conor Russomanno

DURING THIS SUMMER'S DIGITAL REVOLUTION EXHIBITION AT LONDON'S BARBICAN MUSEUM, a small brainwave-influenced game sat sandwiched between Lady Gaga's Haus of Gaga and Google's DevArt booth. It was Not Impossible Labs' Brainwriter installation, which combined Tobii eye tracking and an OpenBCI Electroencephalography (EEG) device to allow players to shoot laser beams at virtual robots with just eye movement and brain waves. "Whoa, this is the future," exclaimed one participant.

But the Brainwriter is designed for far more than just games. It's an early attempt at using Brain-Computer Interface technology to create a comprehensive communication system for patients with ALS and other neurodegenerative disorders, which inhibit motor function and the ability to speak.

The brain is one of the final frontiers of human discovery. Each day it gets easier to leverage technology to expand the capabilities of that squishy thing inside our heads. Real-world BCI will be vital in reverse-engineering and further understanding the human brain.

Though BCI is in an embryonic state — with a definition that evolves by the day — it's typically a system that enables direct communication between a brain and a computer, and one that will inevitably have a major impact on the future of humanity. BCIs encompass a wide range of technologies that vary in invasiveness, ease of use, functionality, cost, and real-world practicality. They include fMRI, cochlear implants, and EEG. Historically, these technologies have been used solely in medicine and research, but recently there's been a major shift: As the technology becomes smaller, cheaper, and woven into the fabric of everyday life, many innovators are searching for real-world applications outside of medicine. It's already happening, and it's often driven by makers.

The field is expanding at an astounding rate. I learned about it two and a half years ago, and it quickly turned into an obsession. I found myself daydreaming about the amazing implications of using nothing more than my mind to communicate with a machine. I thought about my grandma who was suffering from a neurodegenerative disorder and how BCIs might allow her to speak again. I thought about my best friend who had just suffered a severe neck injury and how BCIs might allow him to walk again. I thought about the vagueness of attention

CONOR RUSSOMANNO is co-founder and CEO of OpenBCI and comes from a mixed background of art, engineering, and science. He has dedicated the first chapter of his career to interfacing the brain, rethinking business, and turning crazy ideas into reality.

1. Conor wears an early prototype of the OpenBCI 3D-printable EEG Headset.

2. Russomanno (left) and Murphy demonstrate how to get started with OpenBCI.

3. UCSD researcher Grant Vousden-Dishington, working with OpenBCI at NeuroGaming 2014.

4. An EEG brain map from the OpenBCI software brain wave visualizer.

5 OpenBCI 3D-printed EEG headset prototypes.

6. The latest version of the OpenBCI board.

1 Zachary Tyler Newton

OpenBCI/Conor Russomanno

5 Alex Marcelo

2 Adam Sabh

3 Alex Marcelo

6 Adam Sabh

disorders, and how BCIs might lead to complementary or even supplementary treatments, replacing overprescribed and addictive medications.

I went on to found OpenBCI with Joel Murphy as a way to offer access to every aspect of the BCI design and to present that information in an organized, collaborative, and educational way. I'm not the only one who sees the potential of this amazing new technology. But creating a practical, real-world BCI is an immense challenge — as the incredibly talented Murphy, who designed the hardware, says, "This stuff is really, really hard." Many have attempted it but none have fully succeeded. It will take a community effort to achieve the technology's potential while maintaining ethical design constraints. (It's not hard to fathom a few not-too-far-off dystopian scenarios in which BCIs are used for the wrong reasons.)

Of the many types of BCIs, EEG has recently emerged as the frontrunner in the commercial and DIY spaces, partly because it is minimally invasive and easily translated into signals that a computer can interpret. After all, computers are complex electrical systems, and EEG is the sampling of electrical signals from the scalp. Simply put, EEG is the best way to get our brains and our computers speaking the same language.

EEG has existed for almost a hundred years and is most commonly used to diagnose epilepsy. In recent years, two companies, NeuroSky and Emotiv, have attempted to transplant EEG into the consumer industry. NeuroSky built the Mindwave, a simplified single-sensor system and the cheapest commercial EEG device on the market — and in doing so made EEG accessible to everyone and piqued the interest of many early BCI enthusiasts, myself included. Emotiv created the EPOC, a higher channel count system that split the gap between NeuroSky and research-grade EEG with regard to both cost and signal quality. While these devices have opened up BCI to innovators, there's still a huge void waiting to be filled by those of us who like to explore the inner workings of our gadgets.

With OpenBCI, we wanted to create a powerful, customizable tool that would enable innovators with varied backgrounds and skill levels to collaborate on the countless subchallenges of interfacing the brain and body. We came up with a board based on the Arduino electronics prototyping platform, with an integrated, programmable microcontroller and 16 sensor inputs that can pick up any electrical signals emitted from the body —

including brain activity, muscle activity, and heart rate. And it can all be mounted onto the first-ever 3D-printable EEG headset.

In the next 5 to 10 years we will see more widespread use of BCIs, from thought-controlled keyboards and mice to wheelchairs to new-age, immersive video games that respond to biosignals. Some of these systems already exist, though there's a lot of work left before they become mainstream applications.

This summer something really amazing is happening: Commercially available devices for interfacing the brain are popping up everywhere. In 2013, more than 10,000 commercial and do-it-yourself EEG systems were claimed through various crowdfunded projects. Most of those devices only recently started shipping. In addition to OpenBCI, Emotiv's new headset Insight, the Melon Headband, and the InteraXon Muse are available on preorder. As a result, countless amazing — and maybe even practical — implementations of the BCI are going to start materializing in the latter half of 2014 and into 2015. But BCIs are still nascent. Despite big claims and big potential, they're not ready; we still need makers, who'll hack and build and experiment, to use them to change the world. ◉

★ 3D PRINT YOUR
OWN ACCESSORIES

★ CONTROL HIM WIT
YOUR SMART PHONE!

★ ARDUINO-
COMPATIBLE! ★

✴ ROCKET-LAUNCHING ARM ✴

TINKERING TOYS

Illustrated by Matthew Billington

To a maker, a toy can be many things. It can refer to an entertaining product that delights while educating, like Lego sets and electronics kits. It can be an apparatus like a plastic injection-molding machine that helps us produce figurines and widgets. Or it can be a new creation that we produce either for ourselves or as a part of a larger business for others to enjoy. In the following pages we'll help you find the best maker toys, show you how to build toy-making tools, learn from the pros on how to take your toy and game ideas to market, and more. Ready for fun? Then keep on reading.

Games, Gadgets, and Gizmos

SCIENCE

CHEM C3000: CHEMISTRY EXPERIMENT KIT
$250, thamesandkosmos.com

Corrosive and flammable warning signs on the box? You just know it's fun. Whether you're a student wanting to get a head start on high school chemistry — or just chemically curious — this educational kit is never boring. In no time you'll be wearing your safety glasses and taking selfies with your chemistry set-up.

OSCILLOSCOPE KIT
From $45, vellemanusa.com

Buying an oscilloscope is a debate almost every maker with an electronics bent encounters. Forgo the debate and opt to solder this kit. Assembly isn't too tough for anyone with a bit of through-hole soldering experience — and it's also a great way to work on your soldering skills.

Play is key to the growth of young minds, and the right toys can build a child's understanding of advanced concepts and principles to help him or her progress through life with technical confidence.

We want as many new makers to get a good start with their applied education, so we've gathered, tested, and reviewed the best playful products that focus on science, technology, engineering, art, and math. From DIY bots to advanced building blocks, the following list will help give kids — and grown-ups — a valuable grasp of STEAM elements the best way possible — through having fun.

SEEING
$15, exploratorium.edu
Working through a few of this book's 30 hands-on visual discoveries with two 7-year-olds elicited a lot of "whoa!" remarks. Chasing delightful results provided a great incentive to practice observation and reading. For the experiments that involve building, everything we needed was included with the book or easily found nearby.

STRAIN
$25, hungryrobot.com
Real-world biology informs this tabletop game, where players bioengineer microorganism armies complete with organelles and use ATP to build resistance to viruses and release toxins on competitors. But don't worry, it's nonviolent, and there's no chance a pathogen will escape your kitchen table to wreak havoc on your neighborhood.

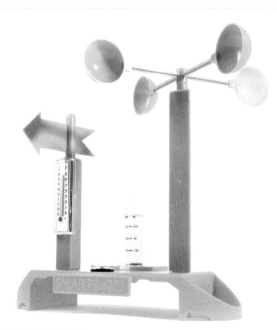

WEATHER LAB
$15, smartlabtoys.com
Rain or shine, this lab will not only get your kids interested in weather patterns, it provides an ongoing project that they can revisit as often and for as long as they like. This easy-to-assemble kit includes a booklet with five experiments and information to help learn how to track wind, chart changes in the temperature, and measure rainfall.

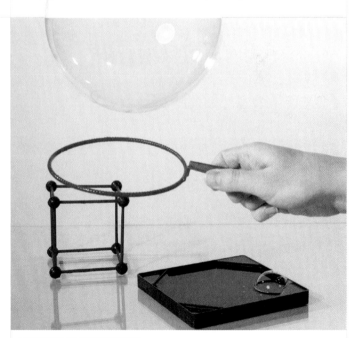

BUBBLE SCIENCE
$35, thamesandkosmos.com
Experiments in this kit go beyond just blowing bubbles, teaching about the science of soap and water that makes bubbles possible. The straightforward, easy-to-read guidebook is packed with recipes, tips, fun facts, and plenty of ideas for a bubble-filled afternoon.

Games, Gadgets, and Gizmos

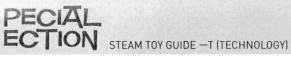

TECHNOLOGY

DARWIN MINI
$500, bit.ly/darwin-mini

Get down with the Darwin-Mini! An Android app (iOS pending) enables wireless operation of this agile, adorable humanoid robot via button, gesture, or voice control. Shipping with a preprogrammed suite of sweet dance and tumbling movements, it uses the sensor-compatible and expandable ARM Cortex M3 OpenCM9.04-C controller board to power 16 highly accurate Dynamixel XL-320 DC servo motors. When you're ready to teach your robot new tricks, it's programmable with the Arduino-like ROBOTIS OpenCM IDE (Macintosh OS X, Linux, and Windows), or the R+ Motion and R+ Task Windows software. STEP and STL files available for 3D-printable customization.

ROBO LINK A
$25, bit.ly/robo-link-a

Who doesn't like to make a cute little robot? With this kit the kids will also learn about mechanical linkage and gears. It is a 5-in-1 kit, and a lot of fun to build. It was, however, slightly difficult to take apart — be careful not to break the Artec Blocks. The instructions are very easy to follow but might be a little tough for younger kids.

MAKE: IT
$80, radioshack.com

The Make: It kit combines a motor shield, two gear motors, two infrared sensors and a host of beams, screws, and nuts. You'll need to supply the Arduino, but it is a great way to rapidly build robots. The kit has options for two moving builds; additional packs are available once those are mastered.

HUMMINGBIRD ROBOTICS KIT
$199, hummingbirdkit.com

This fun little kit gives you the tools you need to imagine and build your first robot. It comes with a solid selection of input and output components, and you can use the kit box for your bot body. The first time we moved our hands in front of the distance sensor and saw our cat-bot tilt his head from side to side, change eye color, and wave his paw at us, it was pure delight.

FLEXBOT
$70, flexbot.cc
Break the airframe on this flyer during a spectacular crash-landing? Don't worry — its design files are freely available online to let you 3D print a replacement; the company also encourages you to create and share new configurations. The supplied instructions are sparse, but very detailed guides are available online.

SPHERO 2.0
$130, bit.ly/sphero-2
This programmable, waterproof, Bluetooth-controlled light-up ball is gorgeously simple (no buttons), and it's that simplicity that makes it so fascinating to watch it roll around. The app design is even more brilliant, with challenges and missions to keep you busy for hours.

HEXY
$250, bit.ly/hexy-bot
Though its body is 80% servo-acrylic technomatter, Hexy is surprisingly adorable and approachable. It's the perfect project for the novice hexapod roboteer — cheap (relatively speaking), easy to build, and easy to run. The Kickstarter runaway success is a true testament of community in the open-hardware movement.

VEX ROBOTICS EDR CLAWBOT
$500, vexrobotics.com
Don't be too intimidated by this; the build isn't as hard as it looks. The fun comes later, when the onboard microcontroller is paired with the remote, and ambitious teens can reprogram the bot with their own code. There are plenty of extra digital and analog I/O ports, more channels to support extra motors, and the company sells a plethora of add-on kits. Be warned, all these features don't come cheap, and the Clawbot is more likely useful in an educational setting. But if your child is really serious, she can get involved in the competitions VEX Robotics offers.

LEGO MINDSTORMS EV3
$350, lego.com
The EV3 is the third evolution of the Lego programmable brick, which handles both hardware connections — motors, sensors, and wireless communication — and software. Programming is done using the On-Brick interface, or by using desktop applications such as Lego's EV3 Software, LabVIEW, or RobotC. It's even possible to control the brick using the company's commander app for mobile platforms. Laudably, Lego open sourced the EV3 software, so keep on the lookout for hacks and community-created software.

ROBOTIC ARM EDGE
$62, bit.ly/robotic-arm-edge
There's only one right way to build this, and lots of little parts to get just so. It'll take some time and some skills — a steady hand and a lot of patience are recommended — but it's good for hours of heads-down focus, and the finished product is something to be proud of.

ROBOTIKITS 6-IN-1 SOLAR KIT
$22, bit.ly/6-in-1
Start by building a gearbox and a solar unit, then combine those with extra parts to make a car, a puppy, a sailboat, and more. The parts go together smoothly and are easy enough to take apart to switch from one project to another. The pieces are small, so the car is adorable; it ran like a champ for us.

Games, Gadgets, and Gizmos

ENGINEERING

ERECTOR MULTIMODELS 25 SET
$40, erector.us

This classic toy is on the short-list of pastimes that are sure to out you as a nerd. But it deserves its geek cred. Yeah, the pieces can be futzy and it takes a little while to get the hang of, but you're using real tools and solving real engineering problems. The best part, though, is dispensing with the directions and designing your own models, especially if you combine multiple sets and lots of universal structural pieces. Standard-size nuts and bolts make parts and kits compatible and interchangeable, but be prepared to invest in some Tupperware to keep all those bits organized.

TREBUCHET KIT
$25, bit.ly/trebuchet-kit

Trebuchets were once the pinnacle of engineering technology, the only machine capable of tossing insanely massive objects over — and through — thick castle walls. This 19-inch trebuchet kit teaches you how the gravity-powered mechanisms work; the finished build can sling a small projectile considerable distances, while adjustable elements show how small tweaks can effect trajectory.

SCOPE CONSTRUCTOR
$60, thamesandkosmos.com

Learn optical science by making up to 28 different versions of telescopes, binoculars, and microscopes with this great kit. The experiment manual is clear and easy to follow, though younger kids may require assistance. One minus: The tube assembly was not a good fit and required tape, which is problematic if you want to reuse tubes. Otherwise, great fun and it works well, too.

PHYSICS WORKSHOP
$55, thamesandkosmos.com

This kit is great for kids interested in the "how" and "why" behind projects and who like figuring things out on their own. Instructions are minimal; the builder figures out assembly for each experiment from just a picture. The manual dives deep into explanations of the topics, including gravity, simple machines, and force. These 37 experiments could keep a kid busy for weeks.

KAPLA BLOCKS
From $40, kaplaus.com

"Better than Legos," one young tester said. That's a hefty claim, but Kapla planks offer an elegant simplicity. Each piece is identical, yet each can serve a totally different purpose. They look like Jenga blocks, and poorly designed builds will act that way too, but with flexible connectors (sold separately) and a little experimentation, kids can create remarkably robust structures.

ROKENBOK ROK WORKS CONSTRUCTION & ACTION SET
$130, rokenbok.com

Sometimes once a toy is built, the thrill is gone. Not so with Rokenbok, which keeps kids driving an RC truck, endlessly loading little blue and red balls onto a conveyor belt and rolling them down the customizable structure they built. Some adult help may be necessary.

DELUXE ROOMINATE
$50, bit.ly/deluxe-roominate

Easy-to-connect electronics — a motor and a light — are the highlights of this "wired dollhouse building kit." Picture-heavy instructions provide plenty of inspiration, as do the craft supplies included in the deluxe version. And while it really makes a room, not a house, our imaginative testers didn't mind one bit.

Hands-on learning and fun with *Make:* magazine's STEAM-based toy guide.

Games, Gadgets, and Gizmos

KINETIC CREATURES: RORY THE RHINO
$40, bit.ly/kinetic-rhino

Your mission: Transform some flat cardboard and a wire into a walking rhinoceros. Ready? Go. This kit gives you everything you need to make this amazing creation, and the coolest part is, you don't need any glue or tape. Popping out the pre-cut and scored shapes for each body part and folding them is satisfying, and a bit meditative. The instructions tell you that the secret to good kit building is, "Be patient and take breaks," and they're right.

BARE CONDUCTIVE GLOWING HOUSE
$27, bit.ly/bare-conductive

Folded paper and conductive ink let you build a simple circuit, but beware the black electric paint — it'll make a mess, and worse, short out the circuit. Some of the stickies came unstuck over time, so a bit of glue — and some coloring tools — will enhance the experience and make the kit worthy of a home in a bigger diorama.

MAGNA-TILES
From $12, magnatiles.com

These magnetic building tiles are an incredibly intuitive way to show children how to bring their 2D doodles into a 3D world. The housing of the magnets allows them to rotate to the best orientation for the strongest connection. The only gripe: The smaller kits didn't have enough tiles to build more complex structures.

GOLDIEBLOX AND THE SPINNING MACHINE
$30, goldieblox.com

The Spinning Machine, best for younger kids, can be built and configured to show how different layouts move the apparatus. GoldieBlox approaches one of the biggest challenges of skill-building toys — engagement — with a fun storyline and a set of characters, keeping girls (and sure, boys too) involved to learn while playing.

EXERGIA CANDLE CAR
$30, bit.ly/candle-car

Sure, you could power a little car with a candle and a turbine. But this kit uses a thermoelectric drive based on a heat gradient between the candle and cold water. The science all takes place at the molecular level. As the atoms heat up, they power the car without the need for moving parts — a technology with the potential to capture waste heat in real-world situations.

MINI STRANDBEEST KIT

$50, bit.ly/strandbeest-kit

This kit offers makers a real connection with Theo Jansen's remarkable Strandbeest design. It's not an open-ended project, but this kit quickly introduces you to the mechanics behind Jansen's engineering feats. Plus, the finished toy delights everyone. Be prepared to decipher pictures, as instructions are currently in Japanese.

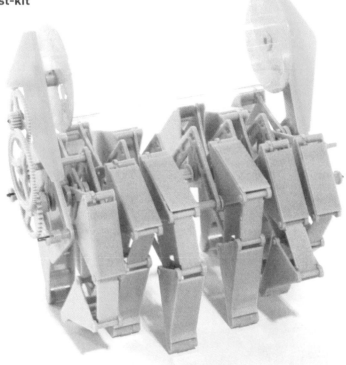

HERBIE THE MOUSEBOT

$40, bit.ly/mousebot

Absolutely nothing is cuter than a PCB board transformed into a boxy looking mouse rolling around the house, maneuvering its way around obstacles with its hypersensitive titanium facial whiskers, searching for the sun using its space-age IR sensors. It's a fun build that will really boost those soldering skills and you'll be proud to say, "Yeah, I made that." As a bonus, every time you add a Mousebot, the cuteness is increased as they chase each other in circles.

K'NEX EDUCATION: EXPLORING WIND & WATER ENERGY

$70, knex.com

K'nex put out these kits specifically so kids could explore STEAM applications with the toys, including wind, water, and solar energy generation. Mechanical power from pouring water into the cups drives a little DC motor that connects to another, which makes whatever else you built go around. That sounds vague, but really, it's an excuse to experiment with drivetrains.

Hands-on learning and fun with *Make*: magazine's STEAM-based toy guide.

Games, Gadgets, and Gizmos

ART

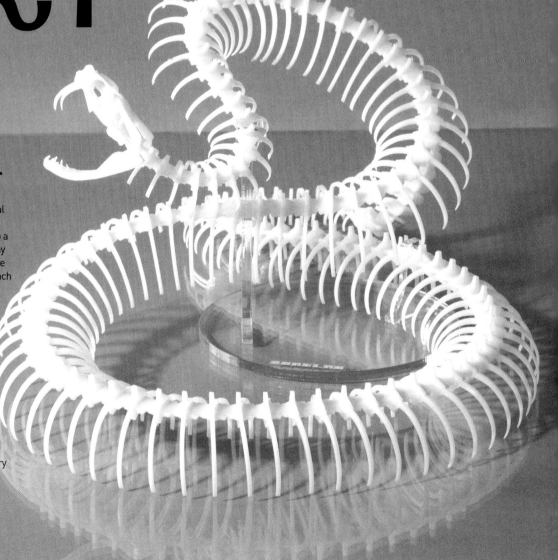

BONELAB RATTLESNAKE KIT

$70, bonelab.com

These laser-cut plastic skeletal modeling kits take your old favorite boxed modeling kits to a whole new level. While you may feel overwhelmed at first by the sheer number of pieces to punch out, the building process goes by much faster and you'll soon forget all about your sore fingers. The pieces fit securely together, and when finished, it is show-cased on the included stand. Pro tip: Don't punch out all the pieces at once. Go by the letters on the sheets and follow the IKEA-like instruction manual. This is a great project for kids and adults who are very patient and detail-oriented.

SUNPRINT KIT

$7, bit.ly/sunprint-kit

Turn ephemeral nature-walk treasures, such as interesting-shaped leaves and flowers, into lasting cyanotype prints. It's easy: Arrange your selections on a sheet of the special UV-sensitive blue paper, wait five minutes while the sun works its magic, rinse in water, and you've made your first Sunprint. This brilliant and super-fun kit brings nature together with an early style of photo printing, and is fun for all ages.

WOOLBUDDY
$24, woolbuddy.com

The instructions were very clear and the process is pretty straightforward. You will poke your fingers with the felting needle, but you'll probably also create something adorable, if slightly deformed. It takes a little practice, but there's enough material for two tries which is really nice. Slightly painful, really fun.

LITTLEBITS KORG SYNTH KIT
$159, bit.ly/littlebits-synth

Learning about pitch, frequency, amplitude, and timbre has never been so much fun, and it's all thanks to the Synth kit. Whether you decide to build the synthesizer, a keytar, or input your own instruments to explore new sonic landscapes, this kit will help you get there.

LIGHTUP MINI KIT
$75, bit.ly/lightup-mini

The Mini Kit has enough pieces to experiment with only the simplest circuits, but includes an app to help tutor and explain exactly what's going on. The magnetic connectors make it easy to swap out components, build in series or parallel, and see how buzzers and LEDs interact with light sensors and variable resistors.

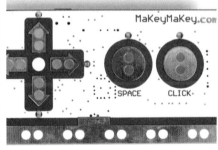

MAKEY MAKEY
$50, bit.ly/makey-makey-kit

Turning a bunch of bananas into a piano keyboard is quintessentially a maker pursuit. The Makey Makey turns common objects into an interface for your computer. With it you can transform a plant into a drum kit, bits of clay into a game controller, or even a stuffed animal into a computer mouse.

ORIGAMI FOR HARMONY AND HAPPINESS
$17, workman.com

Engage in the ancient art of paper folding and create good luck for yourself in the process. This book (paper included) walks you through making symbols and guardians of *feng shui* as well as animals from Chinese astrology. It can be slightly difficult to follow, but the rewards of creation are definitely there. Currently out of print but available used; we also recommend *The Joy of Origami*.

Games, Gadgets, and Gizmos

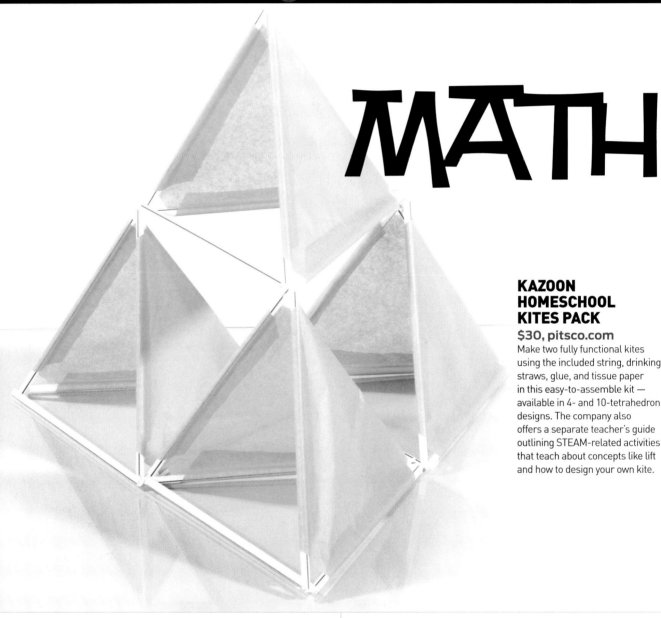

MATH

KAZOON HOMESCHOOL KITES PACK
$30, pitsco.com
Make two fully functional kites using the included string, drinking straws, glue, and tissue paper in this easy-to-assemble kit — available in 4- and 10-tetrahedron designs. The company also offers a separate teacher's guide outlining STEAM-related activities that teach about concepts like lift and how to design your own kite.

SETTLERS OF CATAN
$42, catan.com
Like many Euro- or German-style board games, Settlers favors logical thinking and planning across multiple turns. Manage and trade resources to most effectively expand your domain and earn victory points. Plus, learn probability as players vie for benefits from the most common combinations of dice.

TANGOES
$10, smarttangoes.com
This classic game based on the Chinese Tangram geometric puzzle has gotten no less maddening as it has aged. Play it competitively or just challenge yourself, and when ready for more, go online to download new challenges — or make up your own.

FRABJOUS
$26, bit.ly/frabjous-puzzle
"O Frabjous day!" George W. Hart's sculptural puzzle, named after writer and mathematician Lewis Carroll's nonce word for wonderful, combines 30 curved pieces to define the vertices of a dodecahedron. Both elegant and challenging, each move will have you wondering: Does it go over or under? Buy the puzzle, or cut your own from a template available online.

PERPLEXUS
$20, perplexus.net
This challenging — and addictive — 3D maze can improve fine motor skills while teaching problem solving, patience, and perseverance. One fan confessed he sometimes likes to pretend that the ball is a little car he is driving on the tiny, twisty, colorful roads. Added bonus: self-contained and no batteries required.

VINTAGE MAKER TOYS

VAC-U-FORM
A miniaturized version of a real vacuum forming machine (complete with burnt fingers), this early-'60s gadget from Mattel pulls thin plastic sheets over hard molds for fast reproduction.

CAPSELA
These amazing spheres stand out as our favorite vintage maker toy. EBay a batch, make a dozen different vehicles, and don't be surprised when you fight over them with your kids.

CUBORO
Still available after decades, one 6-year-old tester advises to buy this for "very engineer-inclined" kids who would appreciate that "it has an underground system and an over-ground system."

ROBOT TURTLES
$25, robotturtles.com
Designed to develop a programming mindset in youngsters, this game starts with simple move/rotate options, but quickly introduces additional functions and obstacles to really pull in players. It's addictive — a week after receiving it, a 5 year old was spotted teaching his grandmother to play, and by proxy, how to program.

Doing a Lot with
LITTLEBITS

Written by Mike Senese

In the Manhattan offices of invention company Quirky, the design team responsible for taking a wide range of consumer ideas from mockup to product is busy stacking together a set of interlocking colorful plastic pieces. Working to show new design ideas, the group pulls items from a large set of bins to help test even small variations of their projects. These specific builds aren't made from the famous Danish bricks however, but instead from the magnetically connectable electrical components called littleBits.

"It's already been a hit," says Richard Ganas, one of Quirky's product designers, about the company's new littleBits Pro Library, a massive collection of trim pots, sliders, buzzers, LEDS, motors, sensors, and other gadgets needed to quickly build novel circuits. "We're even tinkering during brainstorms to keep us thinking about what's truly possible to create."
The adoption of the system by serious companies

like Quirky reflects some of the latest moves that littleBits has been making as its product line grows in scope and capability. Developed by MIT Media Lab and Eyebeam alum Ayah Bdeir, the first prototypes popped up online in 2008. The product officially launched at World Maker Faire NYC in 2011 with the 10-piece Starter Kit, which included a pushbutton, slider, and LED bar graph bit. Initial marketing highlighted projects made from paper and cardboard with a fun, simple aesthetic, lending to a kid-friendly educational image for the product — which was not something that Bdeir had in mind as a long-term strategy.

"This was always about creating a tool to democratize hardware and to essentially help hardware go through a revolution the way software did and the way manufacturing did," Bdeir explains, "to really become this creative tool that any person can start with really quickly and innovate ... and invent and create the next big idea."

That next big idea might come from one of the latest bits in the littleBits catalog, the cloudBit. The piece enables wi-fi control of your circuit in various configurations — from the internet to the bit, from the bit to the internet, or from bit to bit.

LittleBits' flexible sensors and Arduino module let this glove's wearer engage in a game of animatronic Rock-Paper-Scissors.

Coupled to a project made from a mix of the now more than 60 littleBit pieces ("There are trillions of billions of combinations," Bdeir proudly states), the new component lets a total novice quickly and easily enter fabled "Internet of Things" territory. Smartphone-controlled pet feeders, a doorbell that sends you an SMS message when pressed, and appliances that adjust their output to environmental conditions are just a few of the early ideas shown with the cloudBit's July launch.

Bdeir sees this new component as a way to use the internet as a building block or material, an alternative to allowing massive companies like Google — which owns Nest — and Apple — which has an upcoming connected home endeavor — prescribe their IOT devices to us. It mirrors her support for the Open Hardware movement, which she helped define as co-founder of the Open Hardware Summit. Bdeir also produced the Open Hardware logo competition, resulting in the now-common gear design that graces hardware that use the open license — including her bits, but not the magnetic connectors, which remain patented to her company. The schematics to those circuits can be found in a Github repository, and Bdeir welcomes people to use them to make their own pieces and learn more about electronics.

Among the other recent littleBits releases is their Arduino module, launched a few months before the cloudBit through the official Arduino at Heart program. It's a simplification of a platform that already simplifies electronics prototyping, and could push both companies to mainstream adoption by eliminating breadboards and finicky components while introducing incredible programmability to a user's creations. Other advanced components are on the horizon as well — Bdeir hints at a camera module and an AC power-tail component, and she says to peek at the idea submissions on their site for other possibilities.

The Arduino release isn't the only partnership with an outside company — over the past year, littleBits worked with instrument maker Korg to release a build-your-own synthesizer kit (see page 43), and with engineers from NASA on a space kit that includes pieces useful for recreating rovers and satellite dishes.

And with the plug-and-play capabilities of the pieces, it's hard to not want to combine them with Lego kits — which Bdeir says helped inspire her creation, alongside the concepts of object-oriented programming. But while a Lego partnership might be a long shot, she is excited about some of the upcoming collaborations.

"There are a couple really, really good ones coming out," she says with a smile. "We like to make a splash." ●

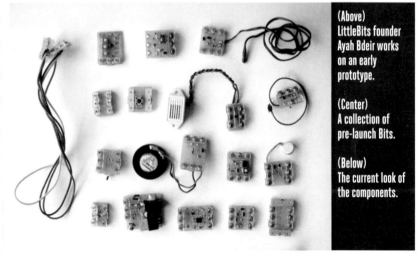

(Above)
LittleBits founder Ayah Bdeir works on an early prototype.

(Center)
A collection of pre-launch Bits.

(Below)
The current look of the components.

Five toymakers share their stories to help you make your idea a reality.

How to Get Your
TOY MADE

So you've got a toy concept and you're sure it will be a hit — that's a good start, but it will probably be the easiest part of going to market. From there you'll have to make prototypes, persuade a toy company to buy or license your idea, or — for those who like to blaze their own trail — find manufacturers and distributors who can produce the toy and get it on shelves. And don't forget about business plans, marketing, and finance.

It's a hard road, but hardly impossible. The following five experts have found success with their toy creations, and have supplied their tips and advice to help smooth a successful toy-making pathway for you.

MAKIE DOLLS — ALICE TAYLOR

Our product is Makies, the doll you make yourself. Makies are (currently) fully 3D-printed, and customized by their owners, who choose and create their own facial features, skin color, eye color, hairstyle and color, clothing — even hand and foot poses.

The idea began in 2010. I drew a few dolls, and sent some sketches to a talented 3D modeler that I found on the Shapeways forum. He modeled up the sketches into 3D using Rhino; we then jointed the doll's limbs, and I sent the final model off to iMaterialise for printing. An 18-inch, bald, eyeless marionette came back, and it cost me 220 euros. That was enough to decide that this could one day be a business.

Since then we've been iterating live with customers. We put Makies live in minimum viable product in mid-2012. We experimented with skin color by boiling all-white printed dolls in tea and coffee. We have Cubes and MakerBots in the office, and we prototype daily on them, printing shoes, jewelry, pets, and more.

The team is now 16 people strong. We use 3D Studio Max, Unity, Solidworks, and Adobe Creative Suite software. We print each doll on-demand and send it direct to the customer's home.

TIPS:

● Always validate your toy idea. Build the smallest, quickest thing you can, and test it with real users: Put it live, point some Facebook ads at it, and get a hundred people you don't know to give you their opinion. Ideally, do this five times with five slightly (or very) different things.

● Always think about your target customers. Know how much they want to spend, why, and when. Know them and their motivations. Why will they care about your (new and unknown) product? Why would they choose your product over something else they know and trust already? Have answers to these questions.

● Do a business plan. Write down your general plan and have a working spreadsheet of numbers that you update and play with regularly. Plan to pivot; plan to evolve. Plan to be agile. Use the numbers as a toy — play with them to see where little changes now can mean big changes later on.

● Work out where the money to survive or disrupt will come from. There's a reason most toys come from only a handful of enormous companies with deep pockets and massive market momentum; they can outspend everyone. They can afford to license expensive brands (Star Wars, anyone?). How do you compete with that?

● Think about your values, ethics, and principles, and how much they'll cost you. It can be very expensive to have principles; it's three to four times more expensive to injection mold locally, compared to having it done in China, or up to 10 times more to use quality, recycled packaging materials. Plan for that.

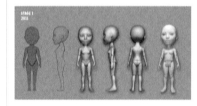

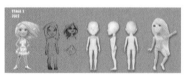

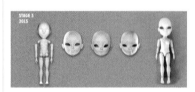

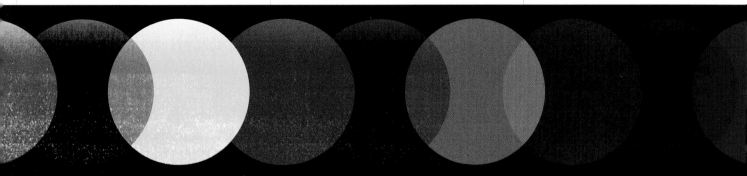

Five toymakers share their stories to help you make your idea a reality.

GoldieBlox

GOLDIEBLOX — DEBBIE STERLING

I created GoldieBlox as a way to bridge the gender gap in STEM fields. The toys introduce engineering concepts through storytelling and building, and kids build alongside Goldie, a girl engineer who solves problems by building simple machines. The first story follows Goldie as she builds a spinning machine to help her dog chase his tail.

I wrote and illustrated it myself, and the prototype was made from wood and materials around my apartment. Once the idea was finalized, we put it on Kickstarter. Within four days, we raised more than $258,000, surpassing our goal and allowing us to start production. We now have several toys available at Toys R Us, Amazon, and more than a thousand retailers nationwide, and we're excited to release three more this month: GoldieBlox and the Movie Machine, GoldieBlox and the Builder's Survival Kit, and our very first action figure of Goldie (who comes with a zip line!).

TIPS:

◉ Make your voice heard. It's the people who stand up for what they believe in and put themselves out there who make a difference.

◉ Make as many connections as you can. Always make time for conversation, and constantly work on establishing new relationships. It's often a second or third connection that ends up being helpful in a major way.

◉ Look for people who believe in your mission as much as you do. When I first debuted Goldie-Blox at the New York Toy Fair, the industry experts told me the idea would never work — they said construction toys for girls don't sell. Within four days of launching, we had 5,000 supporters. That initial fan base has continued to support us and help us grow.

◉ Listen to your customers. One piece of feedback we kept receiving from customers was that kids wanted more — they wanted more pieces to build bigger. So we released GoldieBlox and the Builder's Survival Kit, which has almost 200 pieces.

◉ Get comfortable with being uncomfortable. Starting a new business is like a roller coaster with its peaks and valleys, but don't let the lows bring you down. Learning the ins and outs of running a business takes time, so be open to criticism and change.

DINO CONSTRUCTION COMPANY — BRUCE LUND

Dino Construction Company came out of a conversation more than 20 years ago about favorite toys. Dinosaurs and construction vehicles were two of my most memorable.

We wondered what it might look like if they were combined. We did a couple sketches, including one called TWrex, and showed them to a few companies. They elicited no interest.

We stumbled back upon that sketch every few years and always loved the look. Finally, two decades later, we decided to build one and added a motor, sound effects, and what we called "seek and eat technology." (Top secret, can't say more.)

Many toy companies loved it, but no one needed such a thing. It wouldn't fit in their existing line,or it was too expensive — for one reason after another, it was always a "pass."

We realized that to sell it, we would have to make a line of them, so we built four or five other Dino vehicles, including smaller ones so they'd be less expensive. Still no luck. Our database indicates we showed it 57 times since 2007, the first year we used a database to keep records.

Then one day we had a meeting with a small educational toy company. Someone on our team thought to present them this product, and they loved the look. Go figure. I probably never would have thought to show them that product. They took out all the mechanics and electronics, and made it "kid powered." And kids love them.

TIPS:

• Always give a product a second chance if it is one that you still like whenever you encounter it. We took a sketch and turned it into an item because it was just so cool. We took an item and turned it into a line because the concept had so much potential.

• Always record your ideas in sketch format and file them where you may encounter them again in the future. Had we not made those original sketch illustrations, we would have forgotten the concept long ago.

• Always keep pitching; never give up. Even if you love it, the time may not be right, or maybe you haven't found the right company. It may take five or 10 years, or more. Some of our products it took us more than 15 years to license, before they ultimately became successes, like our game "Doggie Doo" and Dino Construction.

• Always show what we call "wild card" concepts — things that are out of left field, not quite what you think your audience wants to see, because you just never know. I would never have thought that Educational Insights would love our concept and make a vehicle line. And I would have been wrong. But we did, they did, and kids love them.

• Always think bigger, how to make an item into a line of products. Blow up the concept and make it as big as you can. We turned a sketch into a model, and a single model into a line of related concepts.

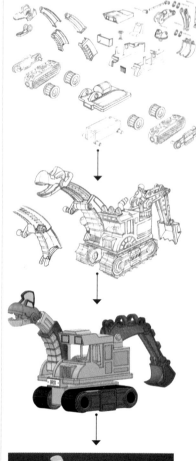

Educational Insights

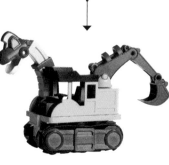

Bruce Lund

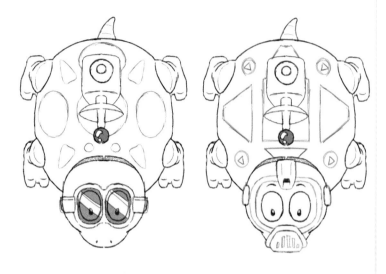

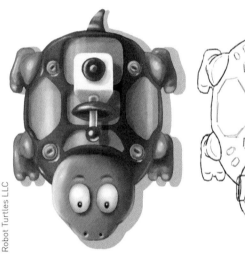
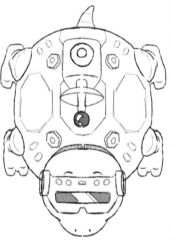

Robot Turtles LLC

Five toymakers share their stories to help you make your idea a reality.

ROBOT TURTLES — DAN SHAPIRO

Robot Turtles is a board game that teaches programming to preschoolers. It didn't start out that way, though. It was born on a lazy weekend last year when I was trying to think of something fun and different to do with my 4-year-old twins. Version 1.0 was inkjet printouts. My kids loved it, so I told friends who thought the game was a great idea. I upgraded to photo paper and a laminator and shared copies with coworkers at Google. Then I decided to get a little more serious, hiring an artist on oDesk, ordering prototypes on The Game Crafter, and rebranding to "Robot Turtles" in honor of the programming language, Logo.

Finally I decided to see if there was enough interest on Kickstarter to warrant a real production run. There was: Robot Turtles became the bestselling board game in Kickstarter history and is now available in stores.

TIPS:

● Sell before you buy. Remember the landfill full of Atari E.T. cartridges that nobody wanted? That's the fate that awaits makers who order inventory before they know the demand. Crowdfunding lets you collect the money first and place your manufacturing order second. Don't let your garage become that landfill.

● Find the hook. I couldn't convince myself that Robot Turtles was a good idea until I came up with the phrase "A board game that teaches programming to preschoolers." That generated interest. It was something new. If I said, "I made an educational board game," nobody would have cared.

● Do the work up front. Don't launch until you know your exact costs for manufacturing, shipping, legal compliance (Robot Turtles had to be tested by a lab for dangerous chemicals), taxes (the IRS watches Kickstarter), and anything else. People bankrupt themselves with successful — but poorly planned — crowdfunding campaigns.

● Budget your time. Whether your launch is a runaway success or a squeaker across the finish line, you'll fill every spare minute answering messages, soliciting bloggers, keeping lines of communication open with your factory, and more. You can launch a product as a hobby — but only if you don't have any other hobbies.

● Be thankful. Whether your sales numbers are 100 or 100,000, every person who supports you is sharing your dream. If people are rude, it's because they care — if they didn't, they'd ignore you. Appreciate the people who are passionate about what you do. We live in an amazing time, where the world can bring our ideas to life. We are lucky makers!

HOG HOLLER — BOB KNETZGER

As an independent toy inventor for the last 30 years, I've worked hard to come up with ideas that could be licensed to toy companies. Here's a case study of a toy idea from blank paper all the way to TV commercial:

Once while listening to a droning lecture my mind wandered. Is there any use for these boring sounds? Maybe I could get the vibration of a sound to make something move. Model train layouts have featured cows that move by vibration and Tudor Games sells vibrating electric football platforms. Could I use the vibration from just the sound of your voice to power a game? I wrote the idea down and made a few sketches while I was thinking about it.

Later, I did some quick plans to work out the design, then went right to a prototype. I built wooden forms, heated plastic in my kitchen oven, and used a Shop-Vac to vacuum-form some plastic parts.

I assembled the parts into a quick prototype, and it worked. By yelling into a tube, your voice would sympathetically vibrate a thin plastic track and propel your "pig" token, skittering along like a vibrobot — but voice powered.

My business partner and I showed the idea to several toy companies — and got rejected. At last, Ohio Art (the Etch A Sketch company) licensed the design and produced it. Hog Holler was on the shelves in toy stores and on TV with a cute commercial. Johnny Carson and Ed McMann even played it on *The Tonight Show*.

TIPS:

● Look for the random connection anywhere. One idea can lead you to another. Even a boring lecture hall can be an inspiration.

● Keep a notebook. Write down and sketch your ideas. (For more on ideation and design sketching, see my article, *"Industrial Design for Makers"* in *Make: Volume 32*.)

● Make a quick prototype of your idea. Fail early and often. You, too, can quickly vacuum-form plastic parts. (See *Make: Volume 11* for my step-by-step DIY article.)

● Keep your beginner's mind. Any really new idea seems silly at first. Don't talk yourself out of it — "Oh, that'll never work. We tried that before. That's not how we do it" — that one crazy idea might just work.

● Don't give up. Be persistent! ◉

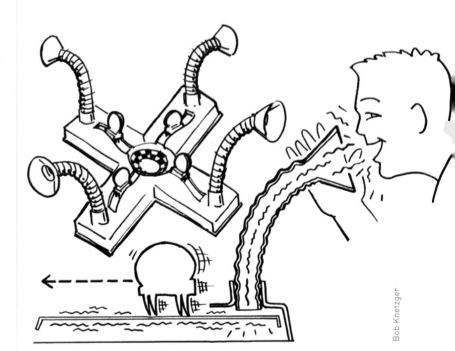

Bob Knetzger

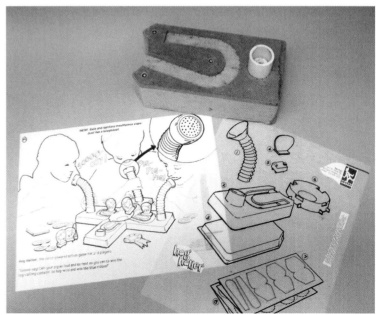

BEYOND BARBIE

These toys can help sow a love of STEAM in girls, simply by doing and making. Use your little one's likes as a starting point and see where it leads.

Empowering girls through play is easy if you start them down the right track.

Flowchart by Anne Mayoral ■ Illustrations by Rob Nance

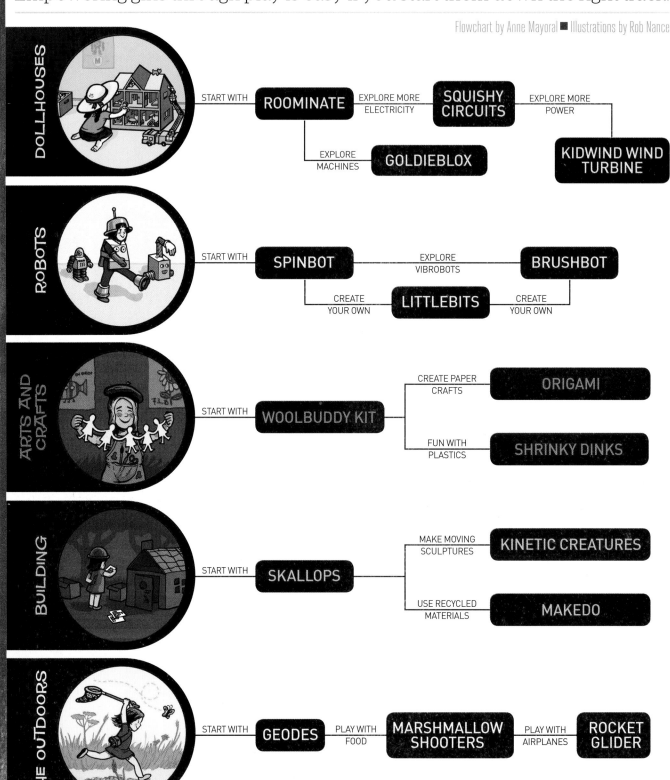

DOLLHOUSES
START WITH → **ROOMINATE** — EXPLORE MORE ELECTRICITY → **SQUISHY CIRCUITS** — EXPLORE MORE POWER → **KIDWIND WIND TURBINE**
EXPLORE MACHINES → **GOLDIEBLOX**

ROBOTS
START WITH → **SPINBOT** — EXPLORE VIBROBOTS → **BRUSHBOT**
CREATE YOUR OWN → **LITTLEBITS** ← CREATE YOUR OWN

ARTS AND CRAFTS
START WITH → **WOOLBUDDY KIT** — CREATE PAPER CRAFTS → **ORIGAMI**
FUN WITH PLASTICS → **SHRINKY DINKS**

BUILDING
START WITH → **SKALLOPS** — MAKE MOVING SCULPTURES → **KINETIC CREATURES**
USE RECYCLED MATERIALS → **MAKEDO**

THE OUTDOORS
START WITH → **GEODES** — PLAY WITH FOOD → **MARSHMALLOW SHOOTERS** — PLAY WITH AIRPLANES → **ROCKET GLIDER**

SODA BOTTLE ROCKET
LED Fireworks

Written by Paul Gentile with Jean Consorti and Lee Siegel

O riginating in ancient China, fireworks have been making people look to the night sky with delight ever since. We wondered: Could we use LEDs to create an aerial fireworks display? The answer is a resounding yes, we can!

During our experiments we created the LED Chutie, which we demonstrated at the 2013 World Maker Faire in New York City. The LED Chutie is an aerial version of the LED Throwie (makezine. com/projects/led-throwies). It uses a 3-volt coin cell battery, an LED, and a plastic bag for a parachute. It's fun indoors too.

1. BUILD A WATER ROCKET

We modified the Soda Bottle Rocket from *Make:* Volume 05 (makezine.com/projects/soda-bottle-rocket) with a bigger nose cone, and made the launcher from instructables.com/id/water-rocket-launcher. This also works well with air.

2. MAKE SOME LED CHUTIES

Slip a 3V coin cell battery between the leads of an LED so that the LED lights up, and tape it in place. Tie a string to each corner of a plastic bag, and tie these to the LED/battery. Fold and roll the parachute neatly but lightly. Repeat.

3. MAKE THE STARS

The center of an aerial firework is called a "star." It's what we see in the sky making all the colors and twinkles that we love. They come in all sorts of chemical compositions, shapes, and sizes.

Our stars are made from LED Chuties and common household items. Place the LED inside a ping-pong ball, plastic golf ball, balloon, or drinking straw to diffuse and enhance the light for different effects. Pack the stars and chutes loosely in the nose cone.

4. LAUNCH!

Prepare several rockets in advance of your show and let the *oooohs* and *aaaahs* begin! ⊘

PAUL GENTILE
(The Hobby Guy) has been making his whole life, from model trains to multicopters. In 2013, he and fellow makers Lee Siegel and Jean Consorti founded Soldering Sunday (solderingsunday.com).

Time Required:
1–2 Hours
Cost:
$20–$60

MATERIALS
» **Plastic soda bottles,** 2-liter (2)
» **LEDs, 10mm** (12)
» **Coin cell batteries, 3V** (12)
» **Plastic bags, about** 12"×12" (12)
» **String, 25'—30'**
» **Ping-pong balls**
» **Plastic golf balls**
» **Latex balloons, small**
» **Drinking straws, large**
» **Foamboard or card-board, about ¼" thick**
» **Masking tape**
» **Cable ties** aka zip ties
» **Rubber bands**
» **Soda bottle rocket launcher** (see step 1)

TOOLS
» **Hot glue gun**
» **Drill with ¼" bit**
» **Scissors**
» **Hobby knife**
» **Epoxy, 5-minute**
» **Marker and ruler**

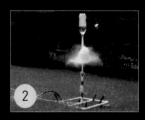

2

3

Get complete step-by-step instructions and photos at makezine.com/projects/soda-bottle-rocket-led-fireworks

Gunther Kirsch

Put the fun in "fun-sized" candy!

Mad Monster
Candy Snatch Game

Written by Bob Knetzger

BOB KNETZGER
is a designer/inventor/musician whose award-winning toys have been featured on *The Tonight Show*, *Nightline*, and *Good Morning America* (see page 53).

Here's a classic toy reimagined for you to make just in time for Halloween candy-giving and party fun. It's the Mad Monster Candy Snatch game, which combines the nerve-wracking dexterity of the old classic Operation game (*BZZZZZT!*) with a fun monster head–shaped candy dispenser. Make those little goblins earn their treats with this tricky game!

It's simple to make and you can customize the play to be as easy or difficult as you like. You can even personalize it with your own voice, choice sayings, and sound effects.

The see-through green monster head is filled with fun-sized candies. Do you dare to snatch a snack? Use the forceps to carefully reach inside its mouth. If you can maneuver out a candy, you've won a treat! But be careful — if you touch the side you lose! The monster wakes up with crackling, shocking sound effects and announces "YOU MAKE MONSTER MAD! YOU LOSE!" as his angry eyes flash red. No treat for you!

1. MAKE THE MONSTER'S HEAD

Empty the soda bottle and save the cap. Rinse it out and remove the label. Cut off the bottom of the bottle and cut points along the opening to form the monster's spiky hair; don't be too neat, he's a mess! Mask off the remaining part of the bottle (again, don't be too precise). Paint the jagged tips of the bottle with black spray paint. When dry, bend each of the triangular points outward to make the pointy "hair."

Go to the project page online at makezine. com/projects/monster-candy-game to download the cutting template (Figure 1). Print it on plain paper. Cut it out on the thick black dashed lines, and fold it over on the thin fold lines. Wrap the pattern around the bottle and using a black permanent marker, trace the cut lines onto the bottle. Use a hobby knife to carefully poke a starting slot. Using a sharp scissors, cut along lines. Punch out the eyeholes with the 3/16" punch to fit the LEDs. Fold the ears and neck bolts at 90° so they stick out.

Cut a piece of aluminum tape 1/2"×6" and cut small slits 1/4" apart all along one long side. Then

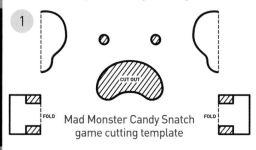

1

FOLD **FOLD**
CUT OUT

Mad Monster Candy Snatch game cutting template

2

Bob Knetzger

cut more slits on the other long side, alternating the cuts so you don't snip the strip all the way through. Then stick the tape to the inside edge of the mouth hole: Place the uncut center part of the tape along the edge, folding over onto the outside and inside of the bottle. It should create a foil-lined edge all along the mouth opening. Cut more pieces of aluminum tape and stick to both sides of the ears and neck bolts.

To finish the head, go to the project page online and download the face label (Figure 2). Print it on an adhesive label sheet and cut it out along the dotted line, being sure to cut out the eyes and mouth too. Carefully center it over the holes on the bottle and adhere it to the outside.

2. MAKE THE BASE

Cut the wood panel to size. I painted mine yellow.

Remove the rubber tip from the doorstopper and find a round-head wood screw that will be a good, tight fit inside the end of the spring. Drill a hole in the center of the bottle cap just big enough for the screw. Thread the wood screw from the inside of the cap and tighten it into the spring.

Drill a pilot hole and fasten the doorstop's mounting base into the center of the wood panel. Twist the spring until it fits tightly. Twist the upside-down bottle onto the cap. Test the spring by filling the monster head with a couple handfuls of candy and giving it a push. The candy-filled head should deflect and wobble but not bend over.

3. MAKE THE CIRCUIT

First, test the sound module: Add a 9V battery then press and hold the REC button and speak clearly into the microphone. Let go of the button to stop recording. You can record up to 20 seconds of sound. If you like, go to the project page online to download and play the sound prerecorded just

Jeffrey Braverman

Time Required:
1–2 Hours
Cost:
$10–$20

MATERIALS

» **Soda bottle, large, green-colored** I used a ginger ale bottle
» **Doorstop spring** Get the kind that has a tapering large-to-small conical shape for just the right amount of bendiness.
» **Aluminum tape** not silver-colored duct tape — real metal tape!
» **Alligator clip jumper wires (6)**
» **Tweezers, long**
» **Knife switch, DPDT** RadioShack #275-1537
» **Sound recording module** RadioShack #276-1323
» **Wire**
» **LEDs, super bright, red, (2)**
» **Power transistor, TIP31 NPN**
» **Resistor, 220Ω**
» **9-volt battery clip with leads**
» **Perf board, small piece**
» **Foam mounting tape, double-sided**
» **Wood board, about 6"×10"×1" thick** for the base. Anything will work: particleboard, plywood, or solid wood.
» **Paint, black and yellow**
» **Screws** from your hardware jar
» **Masking tape**
» **Label sheet, adhesive-backed, blank**
» **Cutting template and face label art** Download them for free at makezine.com/projects/ monster-candy-game and print them out.
» **Halloween candies** "fun size" mini candy bars or any wrapped candy you can pick up with tweezers

TOOLS

» **Punch or drill bit, 3/16"** to fit your LEDs
» **Soldering iron and solder**
» **Scissors**
» **Screwdrivers**
» **Drill and bits**
» **Hobby knife,**
» **Marker, black**

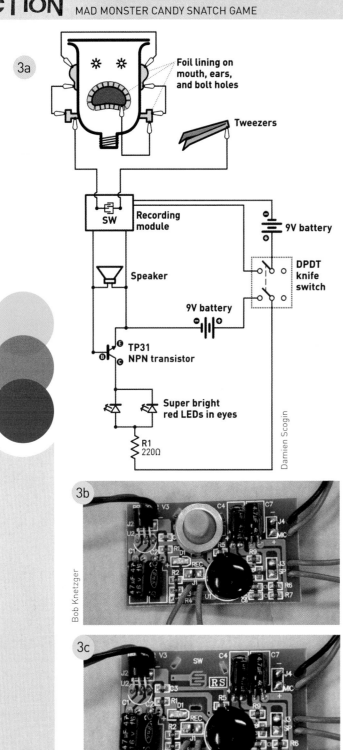

3a Foil lining on mouth, ears, and bolt holes

Tweezers

SW — Recording module

Speaker

9V battery

DPDT knife switch

9V battery

TP31 NPN transistor

Super bright red LEDs in eyes

R1 220Ω

Damien Scogin

3b

Bob Knetzger

3c

3d

for this game. If you prefer, you can instead record your own voice and sounds.

You'll need to slightly modify the sound playing circuit so that the PLAY button contacts are triggered by the game's tweezers and foil sensors. You can see the whole game circuit in Figure 3a.

Carefully remove the PLAY button from the circuit board by prying up the metal tabs on the back of the board. Remove the metal retaining ring and the gray elastomeric button (Figures 3b and 3c).

Insert 2 small wires through the holes on either side of the switch pad (SW) and very carefully solder one wire to each of the 2 traces (Figure 3d). Don't short out the traces! When you touch the ends of the 2 wires together, the sound player should be triggered. Test it!

Solder up the rest of the circuit for flashing the LEDs: Add the power transistor to the perf board and after noting the E, C, and B legs, wire up the connections to the speaker, then solder the connections to the dropping resistor and the 2 LEDs. Wire the LEDs in parallel about 2½" apart on 12" leads that will reach into the monster's head. Wire up the 9V clip, add the second 9V battery, and test the circuit. The LEDs should flash brightly as the sound plays. If not, check polarity on the LEDs, transistor and battery.

Use a small cylinder or plastic cap from a milk bottle to make a resonant chamber for the little speaker. It won't take much to improve the naked speaker's tinny sound. Super-glue the speaker to the cap.

You don't really need the knife switch but it adds a cool "mad scientist's lab" look to the game. Use wood screws to mount the switch to the board. If you want, you can wire the switch to turn the power on and off. Just wire each half of the DPDT switch in series with each of the 9V batteries as shown in Figure 3a.

Use the double-back adhesive foam tape to mount all the components to the board.

4. ASSEMBLE THE GAME

Screw the head back into the spring/cap base. Twist the bottle a little if needed to make the mouth face forward.

Thread the wired-up LEDs through a neck-bolt hole and insert them into the eye holes from the inside. They should fit snugly; if not, tack them with a bit of hot glue or super glue. The wire leads should be loose so they don't restrict the bottle from bobbling.

Now connect everything together using alligator clip jumpers. For an even more mad-scientist look, wrap the wires tightly around a pencil first to give them a "coiled cord" look.

Connect one of the wires from the PLAY sound trigger to the tweezers. Connect the other PLAY

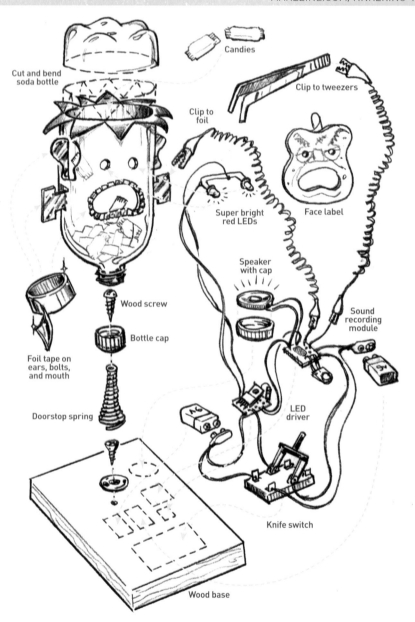

Candies

Cut and bend
soda bottle

Clip to tweezers

Clip to
foil

Face label

Super bright
red LEDs

Speaker
with cap

Sound
recording
module

Wood screw

Bottle cap

Foil tape on
ears, bolts,
and mouth

LED
driver

Doorstop spring

Knife switch

Wood base

Bob Knetzger

trigger wire to the foil on a neck bolt. Use more jumpers to connect the foil mouth, foil ears, and the other neck bolt.

Final test: Close the switch and touch the tweezers to the foil on the edge of the mouth — the monster should talk and flash its eyes! Test the other contact points on the neck bolts and ears with the tweezers, too.

Load up the monster head with some fun-size candies and you're ready to play!

HOW TO PLAY

Easy game: Reach into the monster's mouth with the tweezers — try to get a candy without "waking the monster" (touching the sides). *BZZZZT!* Your turn's over, pass the tweezers to the next player. If you've succeeded, eat your candy or add it to your trick-or-treat bag! You can make the game easier

to win by simply cutting larger holes for the ears and neck bolts.

Simple strategy: Add a die or spinner labeled "Mouth," "Ear," and "Neck." On your turn, spin the spinner and try to snatch a candy from the opening indicated. If you're successful, you can try again, but if you miss you lose all your candies — put them back into the monster's head and let the next player go! Will you risk it — or play it safe?

Name that candy: Player to your right names which specific candy you must try to lift. You may have to do some extra careful digging with the tweezers to win! ◉

Share your monster mods at make-zine.com/projects/monster-candy-game

DIY Rotocaster

Written by
Matt Stultz

Rotocasting is how the pros make hollow plastic parts — and it's easy to do at home.

With a 3D printer you can design and print an endless variety of items and customize them to your needs. But where desktop 3D printers fall down is in making large numbers of the same thing. They tend to be slow and error prone.

One way to quickly manufacture plastic parts in quantity is traditional casting: making molds and pouring resin. But resin can be expensive, and this technique only makes solid models.

That's where rotocasting saves the day. By slowly spinning your mold, you can produce hollow, lightweight plastic parts using one-quarter of the resin. In this project you'll use a 3D printer and a few basic tools to produce your own lightweight rotocaster for a fraction of the cost of a commercial unit.

MATT STULTZ
is the leader of the 3D Printing Providence group, founder of HackPittsburgh, and a MakerBot alum, with experience in multimaterial printing and advanced materials. He wrote "Metal Casting with Your 3D Printer" in our *2014 Ultimate Guide to 3D Printing* and "License Plate Guitar" in *Make:* Volume 37.

1. 3D-PRINT PULLEYS AND GEARS

Download the 3D part files and print them solid at 0.3mm layer height with 100% infill. They're going to take a lot of mechanical wear and tear.

2. MEASURE AND CUT BOARDS

Cut the boards to the lengths indicated in the diagram, using your miter box. Wear safety glasses and keep your hands clear of the blade.

Measure *before each cut*. If you make all the measurements at once and then make all the cuts, your boards won't be the right length. This is caused by what's called *kerf* — the width of the slit cut by your saw blade. On a table saw, kerf is usually 1/8".

3. MARK BOARDS FOR DRILLING

Mark the bearing holes and bolt holes as indicated in the diagram, centered in the face of the boards.

NOTE: NOMINAL 1×3 BOARDS ARE REALLY ABOUT 2½" WIDE, SO MEASURE THE CENTERS CAREFULLY.

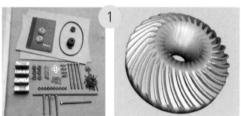

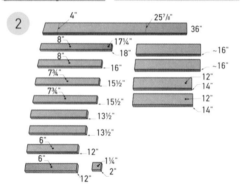

4. DRILL BEARING HOLES

Forstner bits leave clean, flat-bottomed holes — perfect for mounting your bearings. You'll drill the 14" 1×3s, the 15½" 1×2s, the 2" 1×2, and the end hole on the 18" 1×2.

CAUTION: SECURELY CLAMP YOUR WORK BEFORE DRILLING, AND WEAR SAFETY GOGGLES. USE A LARGER SCRAP BACKING BOARD WHEN DRILLING THROUGH-HOLES, SO YOU DON'T CHIP OR "BLOW OUT" THE BACK OF YOUR BOARD.

Place a board on the drill press table and use the Forstner bit to slowly drill a ¼"-deep blind hole at your mark. Test-fit a bearing to ensure it fits flush. Then drill a 7/16" hole in the center of the blind hole, all the way through the board.

5. DRILL BOLT HOLES

Drill the center holes for bolts on the 18" 1×2, the 16" 1×2, and both 12" 1×2s. Slowly drill a 9/16" blind hole 1/8" deep, and test-fit a 5/16" nut in the hole — you want to recess it partially to help lock it in place, but still be able to grip it with a wrench.

Then use the 5/16" bit to drill a through-hole in the center of the blind hole.

6. CUT THE THREADED ROD

Clamp the rod securely and cut it to 9½" using the grinder or hacksaw.

CAUTION: CUTTING METAL WITH A POWER TOOL CREATES SPARKS. WEAR SAFETY GLASSES, LONG SHIRT/PANTS, CLOSED-TOE SHOES, AND LEATHER GLOVES.

7. BUILD THE INNER FRAME

NEW RULE: "LARGE HOLES OUT." DURING ASSEMBLY, ALL BLIND HOLES SHOULD FACE OUTWARD, THROUGH-HOLES INWARD.

Use your corner clamps to build a square frame with the 12" boards butted inside the 13½" boards. Fasten each joint with wood glue and two 1½"

Matthew Stultz

Time Required:
A Weekend
Cost:
$40–$80

MATERIALS

» **3D-printed parts:** 18- and 22-tooth pulleys, right- and left-handed hypoid bevel gears, crank (optional), XL gears (2) (optional) Download the files free from github.com/MattStultz/RotoCaster.
» **Timing belt,** 112 tooth, ¼" wide, 0.2" pitch such as SDI/SP part #A6R3-112025, sdp-si.com
» **Pine board,** 1×3, 8' length
» **Pine boards,** 1×2, 8' lengths (2) You need about 10' total, but get extra.
» **Plywood,** ⅛"×12"×12" (2)
» **Corner brackets,** metal, 3" (4)
» **Skateboard bearings,** 608 type (6) such as Amazon #B002BBGTK6
» **Hex nuts,** 5/16" (10)
» **Lock nuts,** nylon insert, 5/16" (10)
» **Wing nuts,** 5/16" (4)
» **Washers,** 5/16" (4)
» **Threaded rod,** 5/16", 12" length aka all-thread
» **Hex bolts,** 5/16": 3" (2), 3½" (1), and 6" (5)
» **Wood screws:** 1½" and ¾"
» **Screw eyes,** #208, 1⅜" long (4)
» **Cable ties,** 8" (4)
» **Gearmotor,** 10–15 rpm (optional) such as SparkFun #ROB-12115
» **Gearbelt,** XL pitch, 75 tooth, ¼" wide (optional) for motor

TOOLS

» **Miter saw and box**
» **Safety glasses**
» **Corner clamps** (1–4)
» **Combination square**
» **Tape measure**
» **Clamps**
» **Wrench, adjustable**
» **Wrench, ½" fixed**
» **Bit driver** for driving screws
» **Angle grinder or hacksaw**
» **Drill press or hand drill** Use a drill press if you can.
» **Spade bits:** 5/16", 7/16", 9/16"
» **Forstner bit,** 22mm (55/64")
» **Wood glue**
» **3D printer (optional)** Visit makezine.com/where-to-get-digital-fabrication-tool-access to find a printer or service you can use. Or shop for printers at the Maker Shed (makershed.com).

Rotocasting is how the pros make hollow plastic parts – and it's easy to do at home.

8

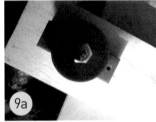

9a

9b

10a

10b

11

12

13a

13b

wood screws. Don't overtighten the screws, as they could split the boards.

8. BUILD THE OUTER FRAME

Build a second frame with the 15½" boards inside and the 16" and 18" boards outside. The 18" board has a 2" overhang at the top.

On the top 15½" board, measure 5⅝" from the overhang and attach the 2" block, oriented with its hole nearer the top than the bottom.

9. BUILD THE BASE

Center the pulley over the through-hole in a 14" board and place the 3½" bolt through (**Figure** 9a). Slide a bearing up the bolt and into the bearing holder. Tighten a hex nut on the bolt, then use two ¾" screws to anchor the pulley in place.

Screw and glue this 14" upright to the 36" 1×3 base board, 4" from the left end, with the pulley facing inward. Attach the other upright 21-⅛" from the first, and secure each with a corner bracket and four ¾" screws (**Figure** 9b).

Finally, add the legs. Center the 2 remaining 1×3s on the ends of the base, attach with screws, then secure with the 2 remaining corner brackets.

10. MOUNT INNER FRAME TO OUTER FRAME

Place the inner frame inside the outer frame. Place a nut in each nut trap in the inner frame and thread a 3" bolt into each hole. Use 2 wrenches to tighten these bolts very snug (**Figure** 10a), but don't crush the wood in the process.

Slide a bearing over each bolt into its bearing holder in the outer frame (**Figure** 10b). Use 2 wrenches to tighten a nylon lock nut on each bolt, until the nuts rest against the bearings and the inner frame is centered in the outer frame. Be careful not to overtighten and bend the frames.

11. MOUNT FRAMES TO BASE

First, set the belt onto the large pulley.

On the outer frame, pass the 6" bolt through the 16" side, and the 3½" bolt through the 18" side, and tighten hex nuts on the outside.

Lift the frame and slide these bolts through the base uprights, with the 3½" bolt going through the pulley. You may need to flex the wood slightly.

Slide a bearing onto each bolt and into its retaining hole. On the 3½" bolt, tighten a nylon lock just flush with the end of the bolt. On the 6" bolt, tighten a nylon lock nut all the way to the bearing, until the nut on the pulley side also rests against its bearing.

12. ATTACH SCREW EYES

Inside the inner frame, screw 2 screw eyes into both the top and bottom boards, spaced about 4" apart and centered.

13. BUILD THE 90° POWER TRANSMISSION

The trick to this rotocaster design is the simple geared transmission it uses to transfer rotation in one axis to another axis that's offset by 90°.

Place a bearing into each of the 2 top holes on the outer frame and slide the 9½" threaded rod through one side. In the center of the rod, tighten 2 hex nuts against each other. Slide the rod through to the other side.

On the outside of the frame, tighten a nylon lock nut 1" down the threaded rod, using the double hex nut as a grip point. Thread the small pulley onto the rod (**Figure 13a**) and put another lock nut on the end to trap it, but don't tighten these lock nuts or the pulley yet.

It's time to set your bevel gears into place. Thread one of them onto the bolt connecting the inner frame to the top of the outer frame. Screw it down tight against the lock nut, then tighten another lock nut on top to grip it in place.

For this next step it's helpful to have an extra set of hands. Screw a nylon lock nut onto the bare end of the 9½" threaded rod, until the bolt just protrudes. Place the second bevel gear against the first one at a 90° angle. Now use 2 wrenches to tighten the rod all the way through the bevel gear, while holding the lock nut. When it's through, use one last lock nut on the end to clamp the gear in between (**Figure 13b**).

Tighten the 2 lock nuts on the other side, so the inner nut is flush against the bearing and the pulley is firmly clamped between the nuts. Now if you turn the rod, the inner frame should spin!

Finally, engage the belt in the teeth of the bottom pulley, and stretch it over the top pulley. It will be tight; if necessary, just get it started on the top pulley and then rotate the device so the belt is rotated the rest of the way onto the pulley.

14. MAKE THE MOLD HOLDER

Measure and cut two 10" squares of ⅛" plywood. I used a laser cutter but you can just use a handsaw. Drill ⁵⁄₁₆" holes through each corner, ¾" from each side.

Slide a washer onto each of your 4 remaining 6" bolts. Stick each bolt through a corner of the bottom plate and tighten a hex nut all the way down. Then slide the top plate about 1" down the bolts and put a wing nut on each bolt (**Figure 14a**).

Use 4 zip ties to loosely attach each of the 4 bolts to the screw eyes on the inner frame. Don't tighten them yet; you'll use them later to balance the mold (**Figure 14b**).

15. MAKE IT MOVE

You need to spin the rotocaster at 10–15 rpm to make it work. There are 3 ways to do this:

» **MANUAL:** I have included a printable crank handle in the 3D files (**Figure 15a**). Thread this onto the 6" bolt and clamp it with 2 lock nuts.

» **SEMI-MANUAL:** Chuck the 6" bolt into an electric hand drill (**Figure 15b**) and lightly press

the button to spin it — slowly! Don't overspin.

» **AUTOMATIC:** Mount a high-torque geared DC motor to your rotocaster. Pair this up with an XL-pitch gearbelt and two 3D-printed XL gears (also included in your free downloads) and it's all set for running fully automated (**Figure 15c**). I power mine with a 12-volt 5-amp laptop power supply and it works great.

USING YOUR ROTOCASTER

To make silicone molds, follow Adam Savage's great tutorial "Primer: Moldmaking" from *Make:* Volume 08 (makezine.com/make-primer-moldmaking-by). I begin by 3D printing the models I want to cast and then sanding and polishing them.

Place your mold into the mold holder and tighten the wing nuts. Spin the rotocaster slowly to check the balance. Tighten the zip ties and adjust the mold's placement in the holder as needed. Then mark the holder where the mold lines up, for future reference.

You're ready to start rotocasting! ⊘

For more photos and tips on resin mixing, rotocasting, and moldmaking, visit **makezine.com/diy-rotocaster**

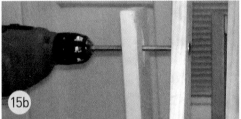

These successfully rotocast piggies contemplate the emptiness within.

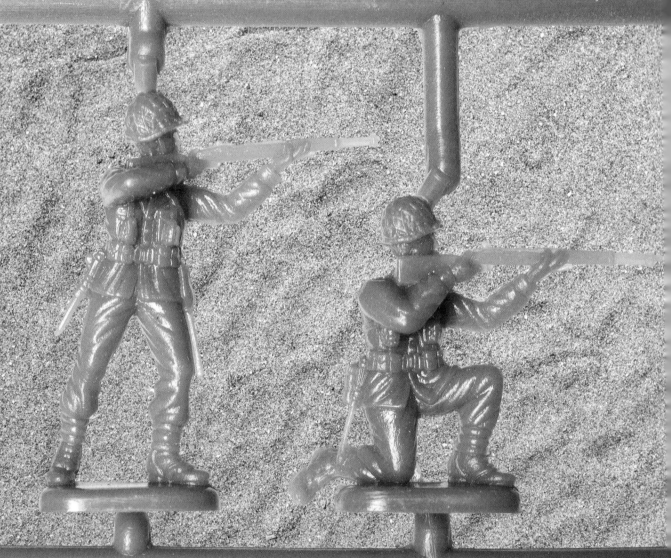

DIY
Injection Molding Machine

Written by Jim Hannon

Tried, true, and totally doable. Start cranking out solid thermoplastic parts.

Classic plastic army men? Yep, they're injection molded.

S ometimes there's a tool you want to buy, but you can't justify the cost. In that case, why not make it? For me, a plastic injection molding machine fell into that category — and it turns out they're not hard to make.

I wanted to make solid plastic parts for some of my amateur science experiments. There are a number of ways to make things out of plastic, each with its advantages and disadvantages. Often just cutting raw material to the desired shape works best. Some

JIM HANNON
retired recently after 39 years as an electri
engineer for Rockwell Collins. Now he can
concentrate on amateur science projects a
on making things in his woodworking and
machine shop — especially things that are
too expensive to buy, or that just don't exist

plastics can be cast by pouring a liquid resin with hardener into a mold (see "DIY Rotocaster," page 60). Vacuum forming works well for making things out of thin sheets of plastic (see "Kitchen Floor Vacuum Former," Make: Volume 11, makezine.com/kitchen-floor-vacuum-former). I considered making a 3D printer, but for the few plastic parts I envisioned needing, it wasn't worth the time and money.

Plastic injection molding has been around since thermoplastics were invented. It's a great way to make many copies of a part quickly, and what I like best is that it's easy to reuse old plastic objects to make new ones.

So I built an injection molder based on the plans in Vincent R. Gingery's book *Secrets of Building a Plastic Injection Molding Machine*. David Gingery could be considered a forerunner of the Maker Movement — he and his son Vincent have written a whole series of books on building tools for the machine shop.

Being an engineer, I couldn't resist making improvements. The plans call for a 1"×1½"×4" piece of cold rolled steel for the heater block Ⓐ, where the plastic is heated before being injected into the mold. I used some 1"×2" hot rolled steel instead. The wider block allowed me to add a second cartridge heater Ⓑ, so my machine warms up quicker and can get hotter.

The frame Ⓒ in the Gingery design is mostly angle and flat iron held together with bolts, but two connections needed to be welded (you could find a friend to do it if you don't have a welder). So I decided to weld most of the frame and avoid drilling so many bolt holes.

The injection lever Ⓓ pivots on a ½"-diameter steel rod Ⓔ. I beefed this up by drilling larger holes in the frame and turning some steel bushings Ⓕ on my lathe to support the rod. The bushings also keep the lever centered over the injection piston Ⓖ.

The major deviation from the plans is the temperature control. In the original design, you have to watch a dial thermometer while fiddling with a bimetal thermostat. Going with something more modern is well worth the effort. I used an inexpensive digital temperature controller Ⓗ from Auber Instruments (auberins.com). These controllers use PID (proportional-integral-derivative) feedback control to bring the temperature up to the exact setting quickly without any overshoot. I mounted mine in a separate enclosure and added a solid-state

Jim Hannon

relay to handle the high-power heaters. A thermocouple sensor Ⓘ comes with the controller; it screws into a tapped hole in the heater block near the nozzle Ⓙ.

Using the machine is easy. The book recommends making a test mold that will make ½" pellets of plastic. These pellets are then used to feed the machine when making real parts. To try it out, I cut ⅜"-wide strips of polyethylene from an old plastic tote lid, set the controller to 380°F, and fed the plastic strips into the cylinder. After the cylinder is filled with sufficient molten plastic, the mold is raised into place under the nozzle. Pull the lever and inject!

Now I'm making test-tube caps Ⓚ for my science experiments (LDPE works well). I'm limited to simple molds I can make with a lathe or mill, but I'm adding CNC capability to my lathe, and thinking about a CNC router. That would open all sorts of possibilities. ◐

+ **Learn more:** *Secrets of Building a Plastic Injection Molding Machine* by Vincent R. Gingery (David J. Gingery Publishing, 1997), ISBN 1-878087-19-3, gingerybookstore.com

⬤⬤⬤ **Get more tips on making and using the DIY injection molder: makezine.com/projects/diy-injection-molding**

Jeffrey Braverman

Time Required:
2 Weekends
Cost:
$100–$200

MATERIALS
» Steel bar, sheet, and angle stock
» Steel rod, ½" and ¼"
» Thermometer, dial
» Thermostat, bimetal
» Cartridge heater, 250W
» Various hardware, wiring

TEMP CONTROL UPGRADE:
» PID temperature controller Auber #SYL-1512A
» K type thermocouple
» Barrier terminal strip, 2-contact
» Relay, 25A solid state
» Switch, SPST, 10A
» Fuse, 3AG, 10A, with panel mount holder
» Extension cord, 3-prong
» Enclosure, 5"×4"×3"
» Wire clamps (2)

TOOLS
» Drill press
» Band saw or hacksaw
» Tap and die set
» Lathe or mill (optional)
» Welder (optional)
» Leather gloves

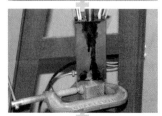

SKILL BUILDER+

MODERATE

BEYOND THE ARDUINO IDE:

AVR USART Serial

Written by Elliot Williams

Time Required:
1-2 Hours
Cost:
$4-$40

ELLIOT WILLIAMS
is a former government statistician and a lifelong electronics hacker.

» He thinks it's hilarious to embed microcontrollers in random electronics projects to give them unexpected capabilities, and he wants you to do the same.

» Before this whole Arduino thing, he taught AVR programming classes at his local hackerspace and even wrote a book based on the experience: *Make: AVR Programming*.

Make:
AVR Programming

Learning to Write Software for Hardware

AVAILABLE IN THE MAKER SHED FOR IMMEDIATE DOWNLOAD: MAKERSHED. COM/PRODUCTS/MAKE-AVR-PROGRAMMING.

Use a $4 microcontroller to launch web pages with the push of a button over serial I/O.

A MICROCONTROLLER IS A SELF-CONTAINED, BUT VERY LIMITED COMPUTER — halfway between a computer and a component.

The top reasons to integrate a microcontroller into your projects are connectivity and interactivity, and one easy way to get your microcontroller talking with the outside world is standard *asynchronous serial I/O*. Many devices can communicate this way, from wi-fi routers to GPS units to your desktop or laptop computer. Getting comfortable with serial I/O makes debugging your AVR programs much easier because the AVR can finally talk to you, opening up a huge opportunity for awesome.

In this Skill Builder, we'll set up two-way communication between an AVR microcontroller and your computer. Your computer will command the AVR to blink an LED, then the AVR will open a web page of your choice in a browser at the push of a breadboarded button using serial I/O.

100101110101...

Asynchronous Serial Communications — A Quick Technical Overview

Computers like to talk to each other in binary: ones and zeros. A simple way to send these binary bits between devices is to connect them together with a wire, and let a high or low voltage on the wire denote a one or zero. Sending bits one at a time like this is *serial* communication because the bits are sent in series, one after the other.

For *asynchronous serial* communication (Figure Ⓐ), there is no common clock signal between the two devices, so to interpret a stream of voltages, each device needs to know how fast the bits are being sent: the *baud rate*.

Atmel's megaAVR family of AVRs (the ATmega series) have a built-in *universal synchronous and asynchronous serial receiver/transmitter* (*USART* for short) hardware peripheral that takes care of all of the hard bits — setting and reading voltages on the serial communication lines at just the right times. Using the USART is not hard: We configure the baud rate, turn the transmitter and receiver sections on, and then feed it data. The USART hardware takes care of the rest.

Configuration

The AVR chips set the baud rate by taking the CPU clock and dividing it down to the right speed. Your main job, in configuring the chip, is to figure out the correct division factor.

The AVR USART hardware samples each bit multiple times to make sure that it's reading a consistent voltage level. In particular, it has two modes: one where it samples 16 times per bit ("normal" mode) and another where it samples 8 times per bit ("double speed" mode). So you're going to divide the CPU speed by 16 and 8 times the baud rate respectively, and then use that result to set the USART timing clock divider.

There are two more catches, though. First, you need to subtract 1 from the result. This is because the AVR's counter starts counting with 0 rather than 1. So if you want the AVR to divide by 4, select 3 for the counter value: 0, 1, 2, 3.

The second catch is tricky, and this is why I think it's worth doing the math by hand as follows at least once. (The AVR standard library includes utilities that help you set up the baud rate automatically.) The AVR only deals in whole numbers, but unless you've chosen your CPU speed to be a multiple of the baud rate, the result of the division above is unlikely to be a round number. So when you round this result off to fit in the AVR, you're introducing some error in the baud rate. A little error is fine, but too much makes the serial communications unreliable.

Baud Rate Example

Say I'm running the CPU at full speed off the internal CPU clock — at 8MHz — and sampling the default 16 times per bit. Let's see which of the standard baud rates work well:

$$8{,}000{,}000 / 16 / 9{,}600 = 52.08 = 0.2\%$$
$$8{,}000{,}000 / 16 / 14{,}400 = 34.72 = 0.8\%$$
$$8{,}000{,}000 / 16 / 19{,}200 = 26.04 = 0.2\%$$
$$8{,}000{,}000 / 16 / 28{,}800 = 17.36 = 2.1\%$$
$$8{,}000{,}000 / 16 / 38{,}400 = 13.02 = 0.2\%$$
$$8{,}000{,}000 / 16 / 57{,}600 = 8.68 = 3.7\%$$

There are a few things to notice here. First, 9,600 baud, 19,200 baud, and 38,400 baud all look pretty good — with the same 0.2% error. The 14,400 baud rate isn't horrible, but around 1% error is pushing it. Rates of 28,800 and 57,600 baud may not work at all with this CPU speed and multiplier combo.

Ⓑ The FTDI Friend, a USB/serial converter. Get it: makershed.com/products/ftdi-friend-v1-0

Parts

» **Atmel AVR microcontroller chip, ATmega168, 168A, 168P, or 168PA series**
» **Breadboard**
» **Pushbutton**
» **Capacitor, 100nF (0.1μF)** to smooth the AVR's power supply
» **LEDs (2)**
» **Resistors, 220Ω (2)**
» **Jumper wires**

Tools

» **AVR ISP (in-system programmer)** with USB cable
» **USB-to-serial converter** such as the FTDI Friend (Figure Ⓑ), with USB cable
» **Breadboard**
» **Power source, 5V DC** if not supplied by your ISP
» **Computer**

Software

» **Project code** Free download from makezine.com/go/avr-usart-serial, includes C code for the AVR, Python code, and a Makefile
» **AVRDUDE and the AVR-GCC compiler** Free downloads for:
  • **Mac:** AVR CrossPack, obdev.at/products/crosspack/index.html
  • **Linux:** Open terminal and type: `sudo apt-get install avrdude avrdude-doc binutils-avr avr-libc gcc-avr`
  • **Windows:** WinAVR, sourceforge.net/projects/winavr/files/WinAVR. Select the option to add AVRDUDE to your path during the install.
» **pySerial package** from pypi.python.org/pypi/pyserial/2.7 for serial communication in Python

MINI PROJECT

Blinking Web-Page-Loader Button

Now you'll breadboard an AVR, program it with an ISP, and connect it via USB-serial so it can communicate with your computer and launch a web page.

1. Breadboard the ISP and Test Power

This example uses the USBtinyISP to provide 5V to the breadboard (**Figure C**). If your ISP doesn't provide power, you'll have to add it.

2. Complete the Programming Circuit and Test Communications

Add the ATmega and the capacitor, resistor, and LEDs (**Figure D**). Attempt to talk to the AVR via your terminal by typing **avrdude -c usbtiny -p m168**. If you get an error, your wiring is incorrect or your chip is bad. Double-check the connections shown in **Figure E**.

3. Flash the C Code to the AVR

In terminal, go to *serialMiniproj*, then compile and load the C program onto the AVR by typing **make**. See hardware setup **Figures D** and **F**.

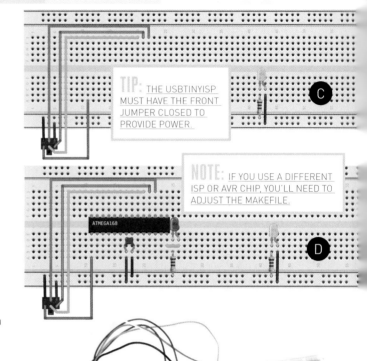

TIP: THE USBTINYISP MUST HAVE THE FRONT JUMPER CLOSED TO PROVIDE POWER.

NOTE: IF YOU USE A DIFFERENT ISP OR AVR CHIP, YOU'LL NEED TO ADJUST THE MAKEFILE.

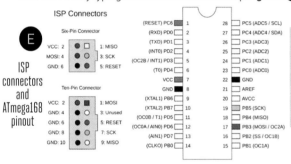

ISP Connectors

Six-Pin Connector

VCC: 2   1: MISO
MOSI: 4   3: SCK
GND: 6   5: RESET

ISP connectors and ATmega168 pinout

Ten-Pin Connector

VCC: 2   1: MOSI
GND: 4   3: Unused
GND: 6   5: RESET
GND: 8   7: SCK
GND: 10   9: MISO

```
(RESET) PC6  [ 1    28 ]  PC5 (ADC5 / SCL)
 (RXD) PD0   [ 2    27 ]  PC4 (ADC4 / SDA)
 (TXD) PD1   [ 3    26 ]  PC3 (ADC3)
 (INT0) PD2  [ 4    25 ]  PC2 (ADC2)
(OC2B / INT1) PD3 [ 5  24 ] PC1 (ADC1)
 (T0) PD4    [ 6    23 ]  PC0 (ADC0)
 VCC         [ 7    22 ]  GND
 GND         [ 8    21 ]  AREF
(XTAL1) PB6  [ 9    20 ]  AVCC
(XTAL2) PB7  [ 10   19 ]  PB5 (SCK)
(OC0B / T1) PD5 [ 11 18 ] PB4 (MISO)
(OC0A / AIN0) PD6 [ 12 17 ] PB3 (MOSI / OC2A)
(AIN1) PD7   [ 13   16 ]  PB2 (SS / OC1B)
(CLKO) PB0   [ 14   15 ]  PB1 (OC1A)
```

NOTE: THE USBTINYISP HAS TWO DIFFERENT CONNECTORS: A 6-PIN AND A 10-PIN (FIGURE E).

AVR Code *main.c*

```c
/* Demo code for serial communication */
/* Demo code for serial communication
 * Sends an 'X' character when button pressed
 * Listens for 'L' and then toggles LED
 * Simple polled-serial style
 * */

#include <avr/io.h>
#include <avr/power.h>
#include <util/delay.h>

// These definitions make manipulating bits more readable
#define BV(bit) (1 << bit)
#define set_bit(byte, bit) (byte |= BV(bit)) // old sbi()
#define clear_bit(byte, bit) (byte &= ~BV(bit)) // old cbi()
#define toggle_bit(byte, bit) (byte ^= BV(bit))

#define LED PB0
#define LED_PORT PORTB
#define LED_DDR DDRB
#define BUTTON PB1
#define BUTTON_PORT PORTB
#define BUTTON_PIN PINB

int main(void) {
  clock_prescale_set(clock_div_1); /* CPU Clock: 8 MHz */

  // Initialize serial
  UBRR0 = 12; /* (8 MHz / 16 / 38400) - 1 */

  set_bit(UCSR0B, RXEN0); /* enable RX */
  set_bit(UCSR0B, TXEN0); /* enable TX */

  // Initialize input/output
  set_bit(BUTTON_PORT, BUTTON); /* set internal pullup resistor */
  set_bit(LED_DDR, LED); /* set output mode for LED */

  while (1) {

    // Poll to see if serial has a byte
    if (bit_is_set(UCSR0A, RXC0)){
     if (UDR0 == 'L'){ /* if the received byte is an 'L' */
      toggle_bit(LED_PORT, LED); /* blink LED */
     }
    }

    // Check to see if button pressed
    // button wired to ground, so a low voltage is a press
    if (bit_is_clear(BUTTON_PIN, BUTTON)){
                        /* wait for send buffer to clear */
     loop_until_bit_is_set(UCSR0A, UDRE0);
     UDR0 = 'X'; /* load up data to be sent */
        /* delay a second to keep from opening too many tabs */
     _delay_ms(1000);
    }
  }                                     /* End event loop */
  return 0;
}
```

The C Code Explained

At the top, we include some standard AVR libraries: **io.h** defines the pins and port mnemonics, **power.h** provides the **clock_prescale_set()** command that we use to instruct the processor to run at full speed, and **delay.h** provides **_delay_ms()** that we use to stall the AVR for a second.

Then we define some convenience macros and the pins that we've hooked up the LED and button to.

Skip down to the **main()** routine and look at the "Initialize serial" section. Calculate the baud rate clock divider as shown in the comments: using the AVR's built-in 8MHz CPU clock at full speed, "normal" mode serial sampling 16 times per bit, and a target rate of 38,400 baud. Doing the division and rounding we get 12; set the USART baud-rate register **(UBRR0)** accordingly. Then enable both the transmit and receive sides of the USART hardware.

The **while(1)** loop is the routine's *event loop*, which runs forever. Inside it, we do two things: Check for data on the serial line, and check for a button press. Within the serial receive code, we directly compare the received byte in **UDR0** with the letter "L". As soon as the USART data register is read, the AVR resets the receive-complete flag, **RXC0**, so that we'll only execute this section of code once per incoming letter.

Finally, if the AVR sees the button pressed, it waits until the USART is ready to send and then loads the data register **UDR0**, with the "X" we wish to transmit. It waits for a second before checking the button again and sending another command. If you leave this bit out, you end up opening multiple web pages before you can pull your finger off the button.

Serial with Python

Now that we've got the AVR speaking serial, it's time to connect it to your desktop, and then on to the world. For the computer-side scripting, I'll be using the Python language because it's simple to learn and lets you do a tremendous amount with very little code.

4. Identify the USB-Serial Port

When using a USB-serial converter, you'll need to figure out how your computer addresses it, and then replace the address **/dev/ttyUSB0** in the Python code below.

On Windows, the USB device will register as a COM port. You can find out which one by looking in Device Manager.

On Linux, the USB-serial device will show up as */dev/ttyUSB0* or */dev/ttyACM0*.

On Mac, open terminal and type **ls /dev/tty.usb*** to see the path. It will show up as */dev/cu.<something>* or */dev/tty.usbserial-<something>*.

Python Code *webLauncher.py*

```
import serial
import webbrowser

sp = serial.Serial("/dev/ttyUSB0", 38400,
timeout = 2)
sp.flush() # clear out whatever junk is in the
serial buffer

  while(True): # endless loop
    response = sp.read(1) # get one byte

    ## Look for open website command
    if response == "X":
      print "Received button press"
      webbrowser.open("http://www.
littlehacks.org/serialEasterEgg.html")

    ## Send blink-LED command
    sp.write('L')
    print "Sending blink command"
```

We import the **serial** and the built-in **webbrowser** libraries, and then open up a serial connection object. We specify the serial port (again, yours may be different) and the baud rate. I've also included the optional **timeout** argument. This is important because otherwise the Python routine will just sit there waiting for data to come in over the serial line.

The other trick here is the **sp.flush()** line — depending on your operating system and what you've connected to the serial port in the past, there may already be data sitting around in your computer's serial buffer. We want to clean that out first, so **sp.flush()** assures that we start with a clean slate.

The Python code is also structured as an infinitely repeating **while** loop. Inside, we read a byte. If there is a byte available, the Python code immediately tests it for "X" and then opens a web page. If there is no byte available within the 2-second timeout, the blink-LED command ("L") is sent.

5. Run the Python Code on Your Computer

You'll need to connect a USB-serial converter to your AVR circuit and add the LED and pushbutton, as shown in Figure **G**. The FTDI cable plugs into your computer's USB port. For two-way (full-duplex) communication, all that's required are two signal wires and a common ground between the two devices to reference the communication-line voltages.

To run the Python program, open terminal, navigate to the *serialMiniproj* directory, and then type **python webLauncher.py** to run the program.

Then with the circuit powered and the USB-to-serial attached to your computer, push the breadboarded button and launch a web page.

That's all there is to it! Two-way communication and control of your AVR and hardware through serial, from your desktop or laptop computer.

Expanding on this simple example is where it gets fun. What can **you** come up with? ✏

G

USB to computer

NOTE: TO FINISH UP THE PROJECT, CONNECT THE LED (THROUGH A 220-OHM CURRENT-LIMITING RESISTOR) TO THE AVR'S PB0, AND CONNECT THE PUSHBUTTON TO THE AVR'S PB1 AND THEN TO GROUND.

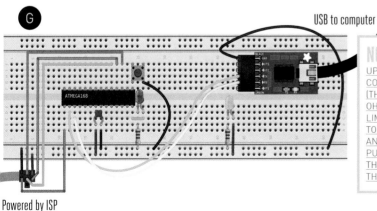

Powered by ISP

EASY

INDUCTION
INSTRUCTION

Written and photographed by Charles Platt

Generate enough power to light an LED through the magic of inductive charging.

CHARLES PLATT
is the author of
Make: Electronics, an
introductory guide
for all ages. He has
completed a sequel,
Make: More Electronics,
and is also the author
of Volume One of
the *Encyclopedia of
Electronic Components*.
Volumes Two and
Three are in preparation.
makershed.com/platt

THERE'S SOMETHING MAGICAL IN THE WAY AN INDUCTIVE BATTERY CHARGER FEEDS CURRENT into a smartphone — or a car — without making an electrical connection.

How does it work? The answer is, the system takes advantage of a fundamental relationship between electricity and magnetism. You can see it for yourself with some simple hands-on experiments.

Electromagnets and Solenoids

I'll begin with that most humble device, the *electromagnet*, which merely consists of some wire coiled around a central rod called the armature. **Figure A** shows a 12VDC electromagnet available on eBay for under $6, capable of lifting about 5 lbs.

A *solenoid* takes this concept a step further. It's an electromagnet with a hollow armature and a separate plunger; which is pulled into the armature when you apply voltage. The plunger usually has a hole drilled at one end, for attachment to an external linkage — such as a lever or an arm. **Figure B** shows an example. You'll need two solenoids to perform the induction experiment that I'm going to describe.

To test it, insert the plunger halfway and apply a 9V battery. *Clunk!* The plunger is dragged all the way in.

Making a Field

Electromagnetism works because when DC current runs through a wire, it creates a magnetic field around the wire. This phenomenon was discovered accidentally in 1820 by Danish physicist Hans Christian Oersted, who noticed a compass needle moving when he applied electric current through a wire.

If you bend the wire into a loop, the magnetic forces combine in one direction to form a *flux*, or magnetic flow, as in **Figure C** (where the flux is visualized in green). This is sometimes called "the right-handed corkscrew rule," because if you turn a corkscrew clockwise with your right hand, it moves downward into the cork. Likewise, if you apply current clockwise through a loop of wire, the magnetic flux pushes downward (assuming we imagine current flowing from positive to negative, and the flux running from north to south).

Now if you create multiple loops in the form of a spiral, you multiply the magnetic force. You can calculate it if you know the number of turns in the coil (*N*), the width of the coil (*W*), and the diameter of the coil (*D*). The approximate relationship is shown in **Figure D**.

(A)

(B)

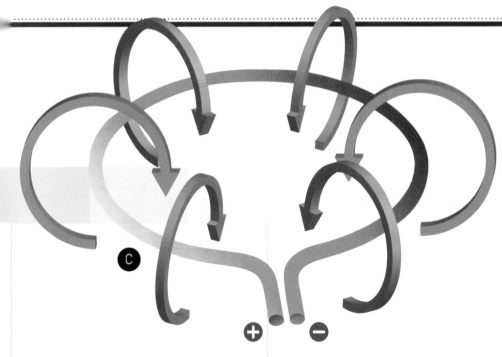

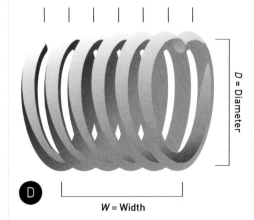

Materials

» **Solenoids, 12VDC, 14Ω coil, about 1.4"×1"×0.7" or similar (2)** from Electronic Goldmine
» **Neodymium magnet, ½" diameter × 1" long** from KJ Magnetics
» **Magnet wire, 2oz spool, 40 gauge or thicker** from Temco Industrial Power (or eBay)
» **Ferrite rods, ¼" diameter × 1¼" long (2)** "33" grade or similar, for low frequencies, from Alltronics
» **555 timer chip, TTL version**
» **Trimmer potentiometer, 250kΩ**
» **Resistors, 4.7kΩ (2)**
» **Ceramic disc capacitors: 560pF (1), 0.1µF (1)**
» **LED, 10mA forward current (or less), 1.6V forward voltage (or less)** Everlight HLMPD155A or similar
» **Battery, 9V**
» **Hookup wire, 26 gauge or smaller, 12"**
» **Dowel, ½"×6"**
» **Screw, flat head**

Tools

» **Soldering iron and solder, low-wattage**
» **Wire strippers**
» **Breadboard**
» **Multimeter**
» **Sandpaper, very fine**

Magnetic Give-and-Take

Just as electricity can create magnetic flux, magnetic flux can create electricity. Almost all the power generation in the world relies on this relationship. Our civilization depends on magnetism.

You can use a neodymium magnet to demonstrate this right on your desktop. (These concepts will be familiar to you if you've read my book *Make: Electronics*, but in a moment, I'm going to take them further.) You need a cylindrical magnet measuring ½" diameter and 1" long, and a spool of magnet wire, which is thin copper wire with a very thin transparent coating.

The spool must have a hole in the middle, ⅝" to ¾" diameter. Also, you must be able to access both ends of the wire. If your supplier hasn't left the inner end poking out, you'll have to rewind the whole thing.

Use very fine sandpaper to remove the insulation from each end. Be careful — the wire is fragile. Now solder some hookup wire of about 26 gauge to the ends of the magnet wire. The solder will only stick where you have successfully removed the insulation. See **Figure E**.

Strip the free ends of the hookup wire, apply a thin coating of solder, and you should be able to push them into holes in a breadboard. Place a low-voltage, low-current LED between them. Now power your LED by moving the magnet in and out of the spool rapidly, as shown in **Figure F** (following page).This is easiest if you attach a wooden handle to the magnet. Use a 6" length of ½" dowel with a flat-headed screw inserted in one end. The magnet will cling to the screw.

N = Number of turns

D = Diameter

W = Width

$$\text{Inductance (approximately)} = \frac{D^2 \times N^2}{(18 \times D) + (40 \times W)}$$

Solder joint

F

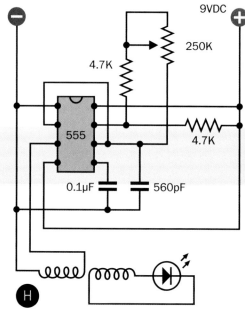

H

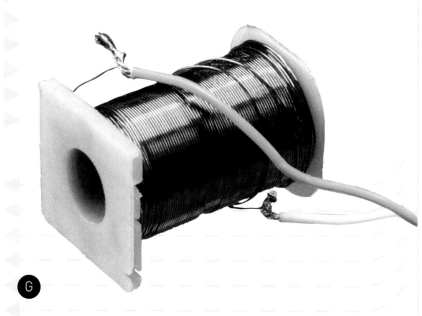

G

This teaches an important lesson. A constant magnetic flux doesn't do the job. It has to fluctuate to *induce* pulses of current. In fact you are generating *alternating current*. Because the LED is a diode, it only responds to pulses in one direction. If you add another LED in parallel with the first but with opposite polarity, they will flash alternately.

To measure the voltage that you're generating, set your meter to AC volts (not DC) and apply it across one of the LEDs.

From Coil to Coil

You've seen that electricity can create magnetic force. You've seen that a fluctuating magnetic force can create electricity. So let's cut out the middleman (the magnet) and put two coils face-to-face. Electricity running through one coil will create magnetism, and the magnetism will induce electricity in the other coil. They will have an inductive connection.

You can use your two solenoids to do this, although you'll have to remove the metal cage enclosing each of them. I pried it off with a hammer and awl to reveal the coil inside, as shown in **Figure G**.

To create a fluctuating magnetic field, the input power must fluctuate. A 555 timer (the old TTL version) can do this, powered with a 9V battery. A simple schematic is in **Figure H**. The trimmer potentiometer will vary the frequency between around 5kHz and 100kHz.

Apply power to the timer, position the coils so that they kiss each other end-to-end, and

check the voltage across the output coil. It will be only around 0.2VAC, because the two coils are not working efficiently. Try adjusting the timer frequency, and you'll see some variation, but not enough to power an LED.

Fantastic Ferrite

You need to channel the magnetic flux with armatures inside the coils. I found that inserting a 1½"×¼" steel bolt in each coil would boost their performance, but not enough. What I needed was a pair of ferrite rods, although you have to get the right kind, because most of them are designed for radio frequencies. A ferrite rod is shown in **Figure**, beside a bolt for comparison. Insert one rod in each coil and place the coils end to end, and you should see your LED light up. If it is very dim, rotate one of your coils 90° relative to the other. Now you should get a bright output. Depending on your particular ferrite rods, some frequencies may work better than others.

Through the magic of inductance, one coil really can power another. A picture of the breadboarded circuit is shown in **Figure** **J**.

Going Further

How can you create the higher efficiency that you find in a real inductive charger? Increasing the frequency, increasing the voltage, using more turns in a coil, or using a larger coil diameter are some obvious options. You can also unwind the wires from the solenoids and wrap them directly around the ferrite cores. Or you can try to wind a pair of very wide, flat coils, which are the type used in charging equipment, and then "tune" the coils with capacitors and other components, although this is not trivial. Go to makezine.com/go/wireless-phone-charge for a tutorial that delves deeper into what's involved. ✏

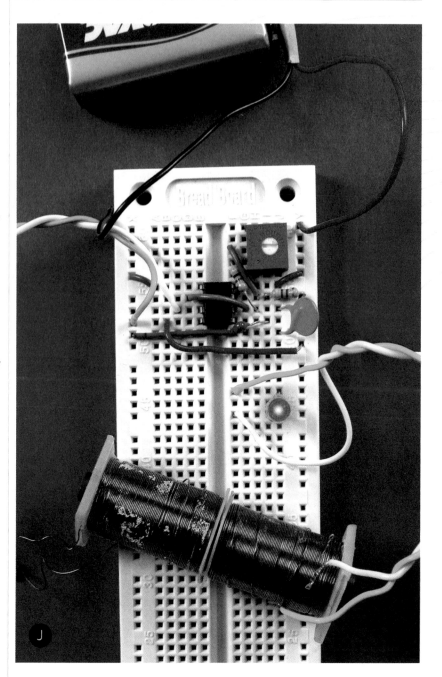

Inductive Charging Bag

Look, ma, no plugs! Make a totally wireless bag to charge your mobile devices, and kiss connectors goodbye.

Written and illustrated by Sean Michael Ragan

**Time Required:
A Long Weekend
Cost:
$100–$200**

SEAN MICHAEL RAGAN
(smragan.com) is a writer, chemist, and longtime *Make:* contributor. His work has also appeared in *ReadyMade*, *c't - Magazin für Computertechnik*, and *The Wall Street Journal*.

LAST NOVEMBER, ADAFRUIT INDUSTRIES' WEARABLES GURU BECKY STERN PUBLISHED A "CELL PHONE CHARGING PURSE" project that used an inductive power transfer link to wirelessly power a hacked handbag from a base station with a built-in transmitting coil. A receiver was sewn into the bag's bottom lining and wired to a phone charging cable inside. Keep your phone plugged into the bag, and the bag on the station, and you could just grab the bag any time and go, confident that you were leaving with a fully charged device.

I liked that idea, but wanted to take it a step further — to eliminate plugs from the system altogether. Here's how I did it.

Materials

FOR THE BAG:

- » **Cellphone or other device with wireless power receiver**
- » **Bag with close-fitting device pocket** You could sew your own bag, but I modified an off-the-shelf bag by Winn Leather that has a pocket made for a smartphone (like my HTC Droid DNA).
- » **Add-on wireless power receiver (2)** for smartphones, e.g., Samsung's Galaxy S4. You want the very thin, flat, flexible kind designed to mount behind a phone backplate.
- » **Wireless power transmitter, low profile** I used the Qifull QT10.

NOTE: I USED **QI** WIRELESS POWER MODULES TO MATCH MY PHONE'S BUILT-IN RECEIVER. BUT **PMA** ("POWERMAT") MODULES WILL LIKELY WORK JUST AS WELL. JUST MAKE SURE ALL YOUR MODULES ARE COMPATIBLE.

- » **Copper wire, 20 AWG stranded insulated, red and black, 24" each**
- » **Wire nuts (4)**
- » **Sewing supplies** including thread to match your bag's design
- » **Eyelets (1–4)**
- » **Heat-shrink tubing**
- » **Styrene sheet, 1mm**

FOR THE BASE:

- » **Acrylic sheet, 0.220", 8"×11"**
- » **AC adapter, 5V, 3A minimum** I used Super Power Supply #B00DHSBFRO.
- » **Wireless power transmitters (2)** I used PowerBot PB1020s.
- » **Machine screws, M3: 12mm (20), 6mm (8)**
- » **Hex standoffs, M3: 12mm (8), 10mm (6)**
- » **Recessed rubber feet, conical, 14mm×8mm (6)** such as Amazon #B007HKFPYM

Tools

- » **Lighter** for heat-shrink
- » **Screwdrivers, small Phillips and flat-head**
- » **Soldering iron, solder, and desoldering tool**
- » **Sewing kit** with seam ripper, needles, thimble, scissors
- » **Hammer**
- » **Eyelet tool and anvil**
- » **Drill and drill bits**
- » **Wire cutter/stripper**
- » **Cutting tool for acrylic** You can use a laser cutter, a plastic saw, or just a utility knife to score and snap.
- » **File, small**

HOW IT WORKS

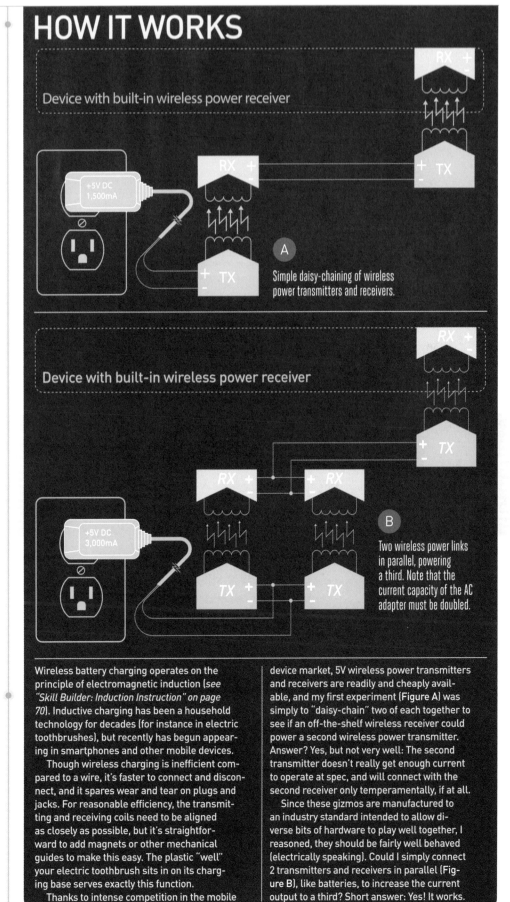

Device with built-in wireless power receiver

A Simple daisy-chaining of wireless power transmitters and receivers.

Device with built-in wireless power receiver

B Two wireless power links in parallel, powering a third. Note that the current capacity of the AC adapter must be doubled.

Wireless battery charging operates on the principle of electromagnetic induction (see "Skill Builder: Induction Instruction" on page 70). Inductive charging has been a household technology for decades (for instance in electric toothbrushes), but recently has begun appearing in smartphones and other mobile devices.

Though wireless charging is inefficient compared to a wire, it's faster to connect and disconnect, and it spares wear and tear on plugs and jacks. For reasonable efficiency, the transmitting and receiving coils need to be aligned as closely as possible, but it's straightforward to add magnets or other mechanical guides to make this easy. The plastic "well" your electric toothbrush sits in on its charging base serves exactly this function.

Thanks to intense competition in the mobile device market, 5V wireless power transmitters and receivers are readily and cheaply available, and my first experiment (Figure A) was simply to "daisy-chain" two of each together to see if an off-the-shelf wireless receiver could power a second wireless power transmitter. Answer? Yes, but not very well: The second transmitter doesn't really get enough current to operate at spec, and will connect with the second receiver only temperamentally, if at all.

Since these gizmos are manufactured to an industry standard intended to allow diverse bits of hardware to play well together, I reasoned, they should be fairly well behaved (electrically speaking). Could I simply connect 2 transmitters and receivers in parallel (Figure B), like batteries, to increase the current output to a third? Short answer: Yes! It works.

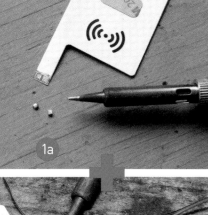
1a

1c

TIP:
ONCE YOU'VE ATTACHED POWER LEADS TO ALL THE MODULES, TAKE A MINUTE TO SET UP AND TEST THE 2-STAGE WIRELESS POWER TRANSFER LINK ON YOUR BENCHTOP TO UNDERSTAND ITS OPERATION AND VERIFY THAT ALL THE PARTS ARE WORKING CORRECTLY.

1b

Instead of a single wireless power link, this bag uses two: one to power the bag from the base station, and a second to charge the phone inside the bag. Build this wireless inductive charging bag and your phone may never need a cord again.

1. Prep the modules

Job one is to solder power and ground leads to all the circuit boards so they'll be easy to interconnect.

1a. The receivers, which are aftermarket add-ons designed to mount between a smartphone battery and backplate, are very thin devices built on flexible PCBs using tiny surface-mount components. Each comes with 2 small gold-plated contact studs to fit the phone's built-in terminals. Remove these by heating them up with your soldering iron, then sliding them off the PCB with its tip. Figure out which pad is power and which ground, then solder 8" of red and black stranded copper wire, respectively, to the pads where the studs used to be. Insulate and secure the 2 solder joints with heat-shrink tubing and electrical tape as needed.

1b. Remove the cases from the 2 base station transmitters and solder 4" red and black stranded jumper wires to their input power and ground connections. These PowerBot "coasters" have handy through-hole-plated test points that make this easy; they're located to either side of the micro USB jack.

1c. Prepare the transmitter that will be sewn into the bag. Since you'll be adding a layer of fabric between it and the receiver, first remove and discard the factory case cover to keep from adding too much material between the two coils. Attach 8" power and ground leads as before. The transmitter I used came with red and black leads presoldered to the main board — all I had to do was desolder and remove the mini USB jack.

2. Install the bag transmitter

Remember that close alignment of transmitting and receiving coils is essential — that's why you need a bag with a dedicated pocket that closely fits your mobile device. If you're sewing a bag from scratch, you can just build in the transmitter board as you assemble the bag. If, like me, you're modding an off-the-shelf bag, you may have to get clever.

2a. Rip out seams as needed to gain access to the bag lining behind the device pocket. Split every second or third stitch with a seam ripper and pull out the little fragments of thread.

2b. Add one or more wiring pass-throughs, as needed, to connect the exposed lining and the bag's main compartment. Each hole should be wide enough to pass your power and ground leads together, and should be reinforced at the edges so it won't tear out in use. Sewing buttonholes is a lot of work, so I opted for hammer-set eyelets. I installed one between the lining and the adjacent side pocket, and one more between the side pocket and the main compartment.

2c. Cut down the bag transmitter's backplate as needed to fit the lining. To prevent snags, smooth off the cut edges and any other protruding case features with a file. Drill a hole in the backplate to provide strain relief, tie a knot in the wires, and pass them through it.

2d. Align the transmitter behind the device pocket and fix it in place. Depending on your bag design, it may be possible to sew the transmitter right where you want it. In my case it really wasn't, so I mounted the transmitter, with cement and short screws, to a piece of thin sheet plastic cut to closely fit the lining. Once the seam is sewn up again, the transmitter won't be able to slip out of alignment with the pocket.

2e. Route the power and ground leads out of the lining via the pass-through.

2f. Temporarily connect a 5V power supply to the transmitter leads and verify that your device charges correctly when inserted into the pocket. Once you're sure the transmitter-pocket alignment is OK, sew up the lining with the transmitter inside, using a needle and matching thread to replace any seams you ripped out earlier.

3. Fit the bag receivers

The receiver modules are small, light, thin, and flexible, and easier to install than the transmitter.

3a. Open the bottom lining of the bag's main compartment. You can rip out a seam, if you like, but it's faster and easier to just make a cut. The repair may be ugly, but it will be hidden inside the bag.

3b. Slip the 2 receivers into the bottom lining and fix them in place with fasteners, adhesive, and/ or stitches. The bottom lining in my bag fit the receiver modules end-to-end almost exactly, so all I had to do was sew it up.

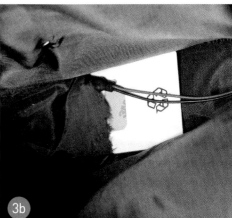

4. Connect the dots

To finish the bag, you just have to connect the wiring between the transmitter and the receivers.

4a. Route the wires from the transmitter into the main compartment through your pass-through(s).

4b. Use one wire nut to connect the 3 red leads together, and another to connect the 3 black leads.

4c. Set up the 2 remaining transmitters on your benchtop and temporarily connect them in parallel to your AC adapter. Put your mobile device in the pocket, then set the bag down on top of the 2 transmitting coils and fiddle with their positioning until all the transmitters and receivers link up and your device starts charging.

4d. Once you're sure the bag's wiring is working, tuck the wire nuts and the receiver "pigtails" into the bottom lining, leaving only the transmitter wires protruding. Sew up the bottom lining around them, and the bag is done.

5. Build the base station

This can be as simple as a flat board with the 2 transmitters glued to its surface. The only really critical factor is that the spacing between the transmitter coils matches that of the bag's receiver coils as closely as possible.

5a. Experiment with the bag and your mobile device to determine the optimum spacing between the 2 base station transmitters. An ammeter connected between the power supply and the transmitters can be useful here; look for the spacing that gives maximum current draw while the bag is sitting on the base with your device charging in the pocket.

5b. Construct a suitable base station with the 2 transmitters fixed at their optimum spacing and securely wired in parallel to the AC adapter. A simple, attractive design — dimensioned for my parts — is illustrated on page 79. It's a good idea to seal the coils with a clear coat of epoxy to protect them from wear.

6. Charge and go!

To use, plug the base station into a wall outlet, put your phone in the bag's device pocket, and set the bag on the base. The Qi standard recommends that transmitters provide auditory, visual, or tactile indication of a successful alignment; usually this takes the form of an LED that changes color and a short beeping sound. Because this design includes 3 separate power transmitters, each of which makes a sound, it's easy to tell just by listening if everything has linked up correctly when you set down the bag. ◗

Share this project online and trade induction charging tips at makezine.com/wireless-charging-bag

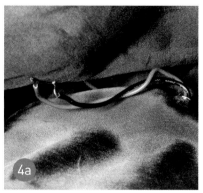

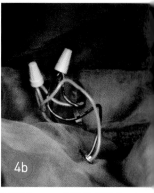

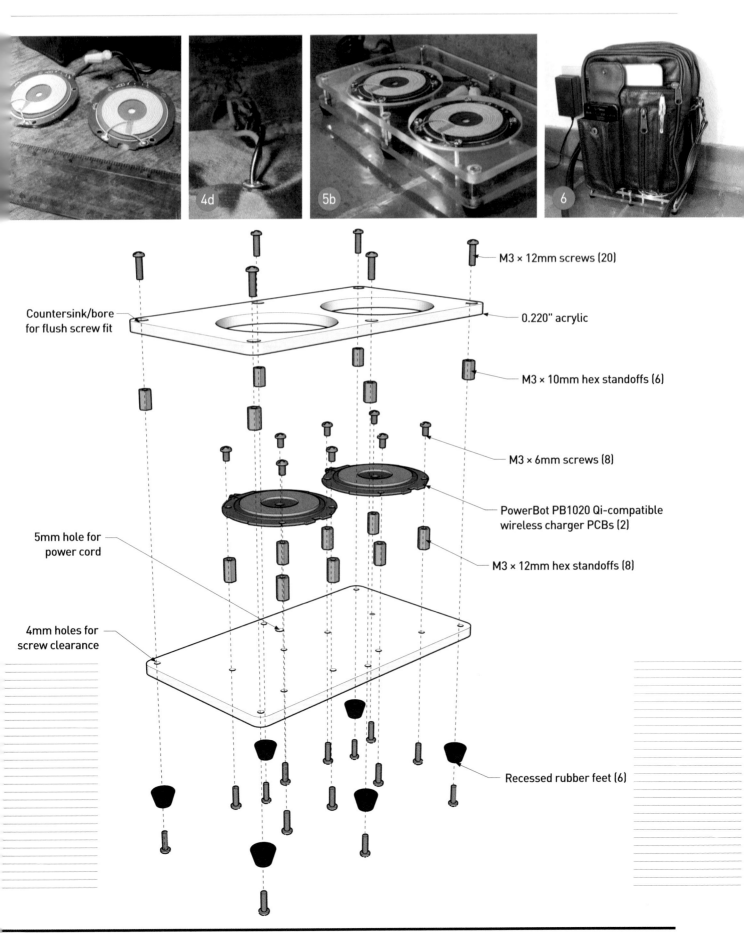

4d

5b

6

M3 × 12mm screws (20)

Countersink/bore
for flush screw fit

0.220" acrylic

M3 × 10mm hex standoffs (6)

M3 × 6mm screws (8)

PowerBot PB1020 Qi-compatible
wireless charger PCBs (2)

5mm hole for
power cord

M3 × 12mm hex standoffs (8)

4mm holes for
screw clearance

Recessed rubber feet (6)

3D-Printed
Pinhole Camera

The fully functional P6*6 camera uses 120 roll film, comes in 35mm and 50mm lengths, and is printable without support even on the tiniest of print beds. *Written and photographed by Todd Schlemmer*

TODD SCHLEMMER
is a firefighter/paramedic living in Seattle. He has studied paleontology and broadcasting, and has worked as a chef, a Birkenstock store manager, and a developer support engineer at Microsoft. His hobbies include 3D printing, photography, writing, computers, robotics, boatbuilding, auto mechanics, ham radio, and cooking. Todd is currently assembling a Shapeoko CNC router kit and lusts for a laser cutter.

THE P6*6 IS A 3D-PRINTED PINHOLE CAMERA, glued and fastened together with 3mm nuts and bolts. All of the individual parts print without support and fit on a 6-inch square print bed. The files are available for download from thingiverse.com/thing:157844.

The P6*6 comes in two focal lengths, 35mm and 50mm. It uses 120 roll film and makes an impressive 6cm square negative — roughly 4 times larger than a negative from a standard 35mm camera. 120 film is widely available and can be found at camera stores that cater to professional photographers or from internet vendors.

1. Print your camera parts

Download the 3D files from Thingiverse and print the camera parts (**Figure ❶**):

- » **A** — Knob, used to advance the film
- » **B** — Cap, snaps onto the body
- » **C** — Baffle
- » **D** — Winder, engages the take-up spool
- » **E** — Film clip, keeps film tightly wound on the spool during unloading
- » **F** — Frame slide, allows viewing of frame number on film backing
- » **G** — Body

P6*6 SPECS:

- » 120 film, 6×6 format
- » 50mm focal length:
  - • f-stop of f/167 with 0.30mm pinhole
  - • 62 degree vertical and horizontal angles of view
- » 35mm focal length:
  - • f-stop of f/135 with 0.26mm pinhole
  - • 77.4 degree vertical and horizontal angles of view

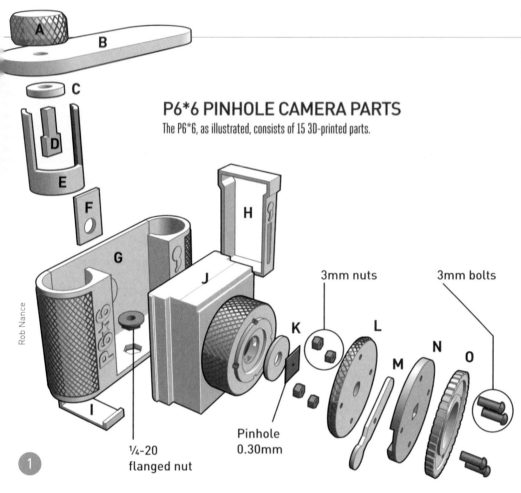

P6*6 PINHOLE CAMERA PARTS
The P6*6, as illustrated, consists of 15 3D-printed parts.

Rob Nance

¼-20
flanged nut

Pinhole
0.30mm

3mm nuts

3mm bolts

①

Time Required:
90 Minutes
Cost:
$12

Materials
» **3D-printed parts** see Step 1
» **Nuts, 3mm (4)**
» **Bolts, 3mm×15mm long (4)**
» **Washers (4) (optional)**
» **Flanged nut, ¼-20** for tripod mount
» **Adhesive-backed velvet or similar** to "trap light" that would leak through joints in the assembly
» **Translucent red plastic, 15mm–18mm disc** A cheap plastic binder is a good source.
» **Thin sheet metal with 0.26mm or 0.30mm pinhole**
» **Black permanent marker** for back of pinhole (no internal reflections!)

NOTE: RELAX — THE PINHOLE IS NOT AS CRITICAL AS IT SEEMS. YOU CAN PURCHASE A PRECISELY LASER-DRILLED PINHOLE ON THE INTERNET OR EASILY MAKE YOUR OWN FROM BRASS SHIM STOCK, A SODA CAN, PIE PLATE, ETC. (ALUMINUM FOIL IS TOO FRAGILE.)

Tools
» **3D printer (optional)** with black ABS or PLA filament other colors will require flat black paint on interior surfaces. To find a machine or service you can use, see makezine.com/where-to-get-digital-fabrication-tool-access
» **Flat files, large and small**
» **X-Acto / hobby knife**
» **Sandpaper, 500-1000 grit**
» **C-clamp or rubber bands**
» **Allen wrenches, small** for bolts, and for manipulating tiny nuts inside the extension when assembling the shutter
» **Drill with ⅛" bit**
» **Epoxy, dark** such as JB Weld
» **Super glue** aka cyanoacrylate (CA) glue
» **ABS plumbing glue, black** for ABS only
» **Plastruct "Plastic Weld"** will bond all manner of plastics

» **H/I** — Body clip and leveling spacer, prints as joined pieces
» **J** — Extension, 50 mm or 35 mm length
» **K** — Pinhole disc, replaceable pinhole mount
» **L** — Pinhole clamp
» **M** — Shutter blade
» **N** — Shutter clamp
» **O** — Trim ring

When preparing the STL files for printing, use the following slicing settings:
» 0.25mm layer height
» 2 perimeters (or "shells")
» 3 solid layers top and bottom
» 50% infill

2. Smooth and fit the printed parts

Every joint between parts in the P6*6 has a potential for photo-ruining light leaks — unintended openings that allow light into the camera. Careful attention to fit will ensure awesome photos. If necessary, use fine sandpaper or a file to smooth mating surfaces.

Carefully enlarge bolt holes with a ⅛" drill bit.

Pay special attention to the frame surface, formed by the bottom of the extension — the film slides across this surface when winding,

The film slides across this frame surface. Any irregularities could cause scratches.

② Make sure this edge is slightly rounded on both sides of this part.

and it forms the margin of your photographs. Surface irregularities could scratch the film, and an unevenly trimmed inside perimeter will be preserved as an uneven border on every photograph you make. Additionally, slightly round and smooth the bevel edge of the frame to avoid scratches on the film (**Figure ②**).

Before proceeding, check the fit of all mating parts. Refer to the exploded parts diagram. All parts should fit together without distorting. The cap should fit the body securely. The shutter blade should be slightly snug between the pinhole clamp and the shutter clamp.

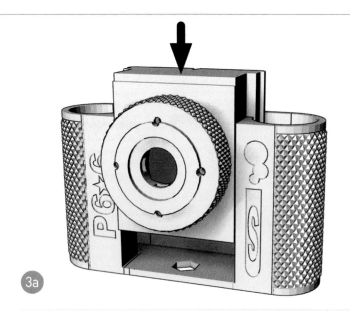

3a

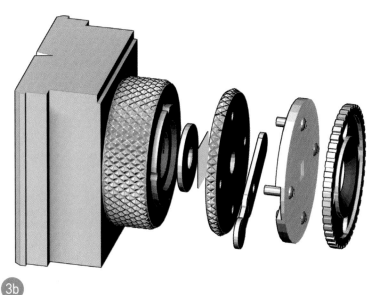

3b

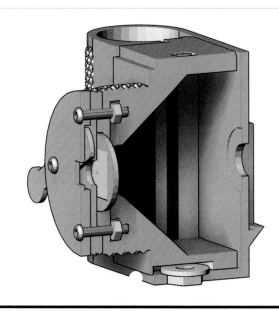

3c

3. Final assembly

TRIPOD MOUNT

A ¼-20 nut is the standard tripod attachment. Carefully bond the flanged nut in its hexagonal hole in the body **(Figure 3a)**, flush with the bottom of the body, using a bit of epoxy on the inside.

CAP AND WINDER

Parts A, B, C, and D (see **Figure 1**).

The winder drive passes through the baffle and cap and into the knob. This is designed to be a friction fit. If the narrow part of the winder drive is slightly too large to fit through the baffle into the knob, enlarge the holes rather than reducing the size of the winder shaft.

EXTENSION AND PINHOLE/SHUTTER

Parts J, K, pinhole, L, M, N, O, nuts, bolts, and (optional) washers. See diagrams.

Everything should fit together tightly prior to fastening. The extension, pinhole clamp, and shutter clamp must fit without interference **(Figure 3b)**.

Bolting all these parts together can be a bit fiddly, but it's important to assemble them before gluing the extension and body together. A small Allen wrench is handy to position the nuts in the nut traps (in the extension) during assembly **(Figure 3c)**. The shutter should snap open and closed. It is easy to overtighten the bolts. Use super glue to mount the trim ring on the face of the shutter clamp.

VELVET LINING AND RED WINDOW

For best results, the inside back of the body can be lined with velvet behind the frame. The velvet provides a gentle friction that keeps the film in place and serves to reduce the effect of stray light from the frame index window. Lining the inside surface of the cap also minimizes light leaks **(Figure 3d)**.

Cut a 15mm–18mm disc of transparent red plastic, and tack it in place in the recess inside the body with a couple tiny dabs of super glue. The hole in the middle of the adhesive-backed velvet will overlap the disc and secure it in place. Carefully use the tip of an X-Acto blade to slide the velvet into position when attaching it to the body and cap. It must be wrinkle free.

BODY/EXTENSION JOINT

"Dry-fit" the extension and body before gluing them together. They'll only fit one way — the "50" (or "35") marking will be visible. Any interference could mean light leaks. The tripod nut must fit without difficulty. Resolve any issues before you glue.

During gluing, space the frame surface about 0.50mm away from the velvet **(Figure 3e)**. You can use 5 sheets of printer paper (0.10mm thick each).

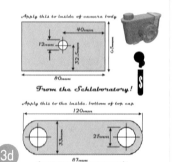

Apply this to inside of camera body
40mm
12mm
32.5mm
80mm

From the Schlaboratory!

Apply this to the inside, bottom of top cap
120mm
33mm
28mm
87mm

3d

Download the self-adhesive velvet dimensions from thingiverse.com/thing:157844

Mind the gaps! A proper fit looks like this.

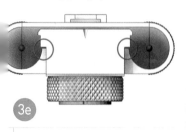

3e

3f

TIP: FOR BEST RESULTS, ASSEMBLE THE SHUTTER AND EXTENSION FIRST.

IMPORTANT: THERE IS A GAP BETWEEN THE EXTENSION AND THE BODY ON EACH SIDE WHEN THE 2 PARTS ARE PROPERLY FITTED (FIGURE 3E).

GLUING

For ABS, plumbing cement works well (and comes in camera black). Work fast — the solvent evaporates quickly and the cement gets rubbery.

For bonding PLA, a dark epoxy is best, but gap-filling super glues or "Plastic Welder" type glues also give good results.

Follow the directions on the label. Too much glue will ooze out of the joint and muck up your lovely camera's appearance. Use a C-clamp or stout rubber bands to precisely clamp the 2 parts together.

Allow the glue to dry, load your new camera with 120 film (Figure **3f**), slide the body clip onto the camera, and make some pinhole photographs (Figure **3g**)!

Get more photos and tips, and share your photos at makezine.com/projects/3d-printed-pinhole-camera

3g

3 FUN THINGS TO 3D Print

Written by Eric Chu

1

2

3

1. Low-Poly Bulbasaur by Agustin Flowalistik
thingiverse.com/thing:327753
This low-polygon representation of Bulbasaur from the Pokémon series really captures the essence of the little monster. Flowalistik is designing more, so follow him to catch (and print) them all.

2. Fuzzy Bear by Robo
(Remixed from Ice Bear by Virtumake**)**
thingiverse.com/thing:71156
A fuzzy, high-polygon texture was applied to this bear using Blender. It provides a tactile experience that feels great in the hand. Try printing with different materials to get different feels.

3. Lolly Box by Faberdasher
youmagine.com/designs/lolly-box
This 3-piece stash box in the form of a popsicle (or ice lolly, as it's known as in the UK and Ireland) is perfect for those sunny days.

LED Concrete Patio Table

Cast the tabletop and built-in cooler in one piece, then make it shine.

Written and photographed by Pete Sveen

PETE SVEEN (aka DIY Pete) enjoys creating projects out of wood, metal, and concrete. In his spare time you can find him whitewater kayaking, flying airplanes, traveling, skiing, and spending time outdoors around Bozeman, Montana. See more of his projects at diypete.com.

Time Required:
2–3 Weeks

Cost:
$250–$350

CONCRETE, LED LIGHTS, AND A BUILT-IN BEVERAGE COOLER. WHAT MORE CAN A GUY ASK FOR? Well, a cold beverage in that cooler I'd suppose!

This concrete-topped table features an integrated concrete trough for ice and beverages and built-in acrylic coasters for your drinks. LED strip lighting shoots through each coaster and runs through the trough too, to light up glass bottles, pint glasses, and wine glasses. The lighting changes colors and is waterproof, so it's the perfect solution to light this table.

The table is very heavy — it took 3 or 4 strong people to flip and move it once it was unmolded — due to the large, reinforced slab of concrete and the extra weight of the trough. I built the trough out of concrete because I thought it would be neat to have the whole thing in one piece, and I had a tough time finding a nice-looking metal trough that was affordable — I didn't want to settle for a plastic planter box. It's built like the high-end concrete sinks you see at fancy restaurants, where the counter is all one piece. Two PVC cap fittings in the bottom of the trough allow for easy draining of melted ice. And you can cover the trough with clear acrylic when it's not in use and still enjoy the lights.

Your guests will be amazed by your new concrete patio table. Set the mood with any color lighting you choose. Or turn on the music and start the dance party! ◕

For step-by-step instructions, materials, tools, and a complete video tutorial, visit makezine.com/projects/led-concrete-patio-table.

1. Insulation foam is used for knockouts below the acrylic coasters. Four tape-wrapped PVC-pipe knockouts create holes in the trough — 2 for stringing the LED strip lights and 2 for drains.

2. Let the concrete cure for 3 to 4 days before unmolding the trough and sides of the table. Wait a full 7 days prior to flipping the table.

3. Polish the concrete with a wet polisher for a super smooth and finished look.

PROJECTS

Try your hand at relief printing, one of the earliest forms of mechanical reproduction.

By Andrew Salomone

ONE-PAGE BOOKS, WHICH OFTEN FEATURE DIY PROJECTS, are typically made from photocopies. Relief printing, on the other hand, is an ancient art. Here's how to combine a modern DIY project with a handmade, linocut block printmaking process to turn your tutorial into a work of art. ◗

Visit makezine.com/onepagebook for the full build tutorial.

Page 1	Front Cover
Page 2	Back Cover
Page 3	Page 6
Page 4	Page 5

Universal CNC
Vacuum Table

Build a vacuum hold-down system with serious suction that works with any CNC router. Written and illustrated by Dan Spangler ■ Photography by Gunther Kirsch

Time Required:
2-3 Days
Cost:
$200-$250

DAN SPANGLER
is the fabricator for *Make:* Labs, and our in-house dastardly moonlight tinkerer.

THOSE OF YOU WITH CNC ROUTERS ARE FAMILIAR WITH THE FRUSTRATION of working with clamps or screws to secure your workpiece and the post-processing required to cut and sand down tabs. A universal vacuum hold-down system can eliminate many of these issues. You simply lay your workpiece on the worktable, turn on the vacuum motor — and voilà! You now have several thousand pounds per square inch of hold-down force securing your workpiece.

Conventional vacuum hold-down systems require you cut a special vacuum template to hold your workpiece tight. *Universal* vacuum tables let you slap your workpiece down anywhere. They need more airflow, as they're comparatively leaky, but they're a lot easier to use.

However, there are some limitations to universal setups. Small parts and layouts with many cuts may not have enough surface area to maintain a strong vacuum, making a universal vacuum table great for most applications — but not all.

The Table Surface:
Base, Plenum, and Bleeder

The universal vacuum table surface consists of 3 layers glued together. First is the *base board*, which bolts to your CNC's frame and attaches

1 Plenum board

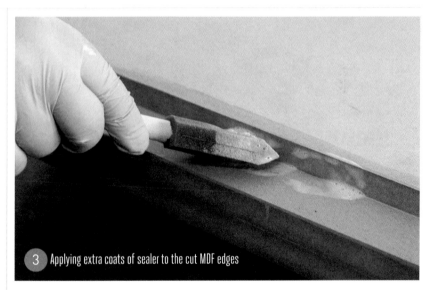

2 Vacuum table air flow

Bleeder board

Plenum

Base board

to the vacuum-system plumbing. Next is the *plenum* board. It contains a grid of airflow channels that distribute the vacuum across the entire table (**Figure 1**). The final layer is the *bleeder board*, a porous spoiler board that allows the air to flow through it (**Figure 2**).

The base board and plenum are typically made from medium-density fiberboard (MDF), as it's cheap and plentiful. (You can also make them out of high-density plastics like ABS or PVC.) These MDF layers are porous and need to be sealed to ensure a strong vacuum. You can use any off-the-shelf wood sealer, such as water- or oil-based polyurethane, epoxy-based sealers, or even wood glue thinned with water (**Figures 3 and 4**). Extra coats of sealer should be applied to the cut edges of the MDF, as they are prone to leaking more than the top and bottom surfaces.

The bleeder board is different. It must be solid enough to support your workpiece but porous enough to allow the air to flow through it relatively unrestricted. The recommended material is Ultralite MDF; it's 40% lighter and more porous than regular MDF due to less epoxy binder. Before you glue the bleeder board to the plenum (**Figure 5**, on the following page), you'll need to shave off the denser top and bottom layers as they can restrict airflow, reducing the vacuum.

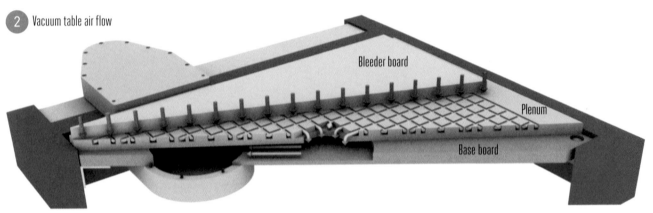

3 Applying extra coats of sealer to the cut MDF edges

The Vacuum

Vacuum systems range from simple, homebuilt setups using Shop-Vacs to commercial units that can cost more than the CNC itself. For most hobbyists and small-production shops, a Shop-Vac or basic vacuum motor (**Figure 6**, on the following page) is more than enough for vacuum table systems.

The goal is to pull both a high vacuum and a high volume of air, as this allows your table

4

Sealing the MDF with wood sealer

ZONES

If you have a large machine, you can divide the plenum into separate zones to improve the efficiency of the vacuum table. Then when you're cutting a small job that doesn't use the entire table you can turn off the zones not in use, gaining more holding power for the zones you are using.

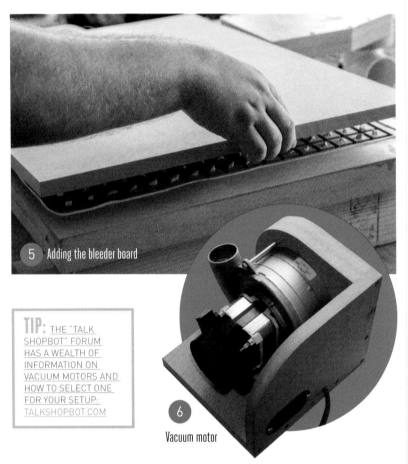

5 Adding the bleeder board

6 Vacuum motor

TIP: THE "TALK SHOPBOT" FORUM HAS A WEALTH OF INFORMATION ON VACUUM MOTORS AND HOW TO SELECT ONE FOR YOUR SETUP: TALKSHOPBOT.COM

7 Vacuum plumbing: good old PVC

to maintain an adequate vacuum despite massive leaks in the system. With a small table area like the 24"×18" surface on the Shopbot Desktop, a typical full-size Shop-Vac works just fine. If you want a dedicated system, you can buy vacuum motors from industrial suppliers like Grainger (grainger.com), or check out Lighthouse Motors (centralvacuummotor.com/lighthouse.htm), which has vacuum motors specifically designed for ShopBot setups.

Using Zones

If you have a larger table and you're dividing the plenum into separate zones, you'll need a way to plumb them and turn different zones on and off. The best solution is basic 2" PVC pipe from your local hardware store (**Figure 7**). Use ball valves to allow control over the different zones, and add a vacuum gauge to the system so you can optimize your airflow and vacuum pressure.

When laying out your plumbing, be sure to avoid lots of sharp turns as these can cause air restrictions, dampening performance. Use Y couplers instead of tee couplers, and align any angled ports to flow in the same direction as the air.

If you find you're not getting enough out of your motor, you can add a second motor, doubling the power. When running more than one motor it's typically better to use 220V power over 110V, as the current draw is less, but that's not always an option. Just be sure that your breaker can handle the current draw.

How to Use Your Vacuum Table

For simple jobs where you have plenty of border around your cuts, just plop your workpiece down on the table, eyeball it square to the edge of the table, then start cutting. It's that easy!

Tips and Tricks
MULTISIDED JOBS

Here's a tip for multisided jobs. When cutting or engraving on both sides of your material, you'll want the cuts to line up perfectly. So, when you cut the first side, incorporate 2 drill holes into your file. Place them in opposite corners on the margins of your material and have the holes extend through the workpiece about ¼" into the bleeder board.

When the first side is finished, stick two ¼"-diameter wood dowels into the holes you previously drilled. Leave ¼" of the dowel sticking up above the surface of the workpiece. Flip your workpiece over, with the dowels still in holes and engage the 2 dowels into the holes in the bleeder board. Your cuts should now line up perfectly.

8 Resurfacing the bleeder board

SHIFTING PARTS

If you're worried about parts shifting while cutting all the way through the material, here's a little trick. On the first pass leave a wafer-thin bit of material (aka "onion skin") at the bottom of the cut to hold the cut parts in place while the rest of the job is being cut. This will maintain a strong vacuum so your previously machined pieces don't shift around.

Once the majority of the material has been removed, go back and run a final pass to cut all the way through. Your first cuts have significantly reduced the cutting force and friction (which decrease with cut depth due to the tiny amount of material left), so there is far less of a chance for your parts to shift.

MAINTENANCE

When you first install your system, note the gauge reading and write it down as a baseline. It should be around 15–20 inches of mercury (inHg) vacuum pressure.

If your bleeder board starts to look rough and your gauge shows decreased vacuum pressure, then it's time to resurface your bleeder board (**Figure 8**).

If you glued your bleeder board to your plenum, you can repeatedly surface your bleeder board until you cut through to the channels in your plenum. Then just glue a fresh bleeder board on top of it. ◗

For a step-by-step build tutorial including materials and tools, visit makezine.com/projects/shopbot-desktop-universal-vacuum-hold-down-system.

3 FUN THINGS TO
CNC

Written by Anna Kaziunas France

1. Wood and Aluminum Bottle Opener
MICHAEL UNA
makezine.com/projects/cnc-bottle-opener-in-wood-and-aluminum
While you're relaxing outdoors playing a game of cornhole, there's no better beverage companion than your trusty, desktop-manufactured bottle opener.

2. Korn-Hole Bean Bag Game
HINES DESIGN LAB
shopbottools.com/files/Projects/POM-KORN-HOLE.zip
Build your own CNC'd flat-pack version of the classic (and much beloved — there's even an American Cornhole Association) bean bag toss game. August Hines' design uses only a single ½" sheet of A/B plywood.

3. Celtic Braid Beer Coaster
BART DRING
makezine.com/projects/solid-oak-celtic-braid-beer-coaster
Now that you've got that drink open, rest it in style on MakerSlide creator Bart Dring's intricate coaster that's small enough to cut on any machine.

Whip It Good: Nitrogen Cavitation in the Kitchen

Infuse foods and drinks with powerful flavor in minutes, using an ordinary cream whipper.

Time Required: 1 Hour
Cost: $60–$100

PALOMA FAUTLEY
is an engineering intern at *Make:* Labs. She is currently pursuing a degree in robotics engineering and has a wide range of interests, from baking to pyrotechnics.

Tools

» **Cream whipper, 1 quart capacity or greater** typically priced at $50–$100
» **Nitrous oxide (N₂O) chargers** aka cartridges or canisters, about $7/dozen
» **Strainer**
» **Cup or glass**

Written by Paloma Fautley

CREAM WHIPPERS ARE USED IN RESTAURANTS TO CREATE PERFECT WHIPPED CREAM INSTANTLY.
But this equipment can be exploited for many other culinary creations, such as rapid flavor infusions and marinades, in a method called *nitrogen cavitation*.

How It Works

Nitrogen cavitation is a method of homogenizing cells and tissues using rapid decompression of gases. Basically, you add nitrous oxide gas (N_2O) to your mixture of ingredients in the cream whipper. The pressurized gas will dissolve into the cells of the ingredients.

When you release the pressure rapidly, nitrogen bubbles form within the cells and expand, breaking the cell walls. This releases flavor compounds quickly, allowing them to easily dissolve in solution and permeate other ingredients. It even tenderizes meats.

The possibilities are really endless, but we have a few recipe suggestions.

Infused Booze

Why steep for a week? Make these recipes in minutes.

LIMONCELLO LIQUEUR

» 4 lemons
» ½–1 cup simple syrup (to your taste)
» Vodka

Slice lemons thinly and discard the ends (these are very bitter). Add lemons and simple syrup to your cream whipper, then add vodka to cover the lemons.

Screw the top onto the cream whipper and add the canister(s) of nitrous, following the manufacturer's instructions to pressurize the whipper. Let sit 10–15 minutes.

Hold the whipper right-side up with a glass over the nozzle to catch any liquid, and release the pressure. Unscrew the top and strain. Enjoy.

SPICED RUM

›› 1"–2" piece of fresh ginger, sliced thinly
›› 1 vanilla pod, split and scraped
›› 2 cinnamon sticks
›› 1 star anise
›› 4 cardamom pods, crushed lightly
›› 2 cloves
›› ½ tsp nutmeg
›› Orange zest (use an unwaxed orange)
›› 1 cup simple syrup, still hot
›› Rum

Add spices and syrup to whipper and let sit 20–30 minutes. Add rum to fill whipper ¾ full. Screw the top on and add the nitrous. Let sit 10–15 minutes.

Hold whipper right-side up with a glass over the nozzle to catch any liquid and release the pressure. Unscrew the top and strain. Enjoy.

Nitro-Marinated Meats

Get overnight flavor and tenderness, right now.

TOTALLY TENDER STEAK

Add about 8oz of meat to the whipper with your de-sired marinade. Small steaks will fit in a 1qt whipper. Larger steaks can be cut into strips or chunks. Screw the top on and add the nitrous. Let sit 15–20 minutes.

Hold the whipper up with a glass over the nozzle to catch any liquid, and release the pressure. Unscrew the top and pour into a bowl. Let the meat sit in the marinade 10–60 minutes, max. Cook as desired.

The cavitation method tenderized our steaks incredibly well. In a side-by-side test it was totally clear which steaks had been put through the whipper. I tested small sirloin portions about 1"×2½"×4" — you get better flavor infusion when you have a lot of surface area — and they came out so tender that it seems a waste to use finer cuts of meat.

PALOMA'S FAVORITE MARINADE

›› ⅓ cup soy sauce
›› ½ cup olive oil
›› Juice of 2 lemons
›› ¼ cup Worcestershire sauce
›› 3 Tbsp dried basil
›› 1 Tbsp parsley, chopped
›› 1 tsp pepper
›› ½ tsp red pepper flakes

Combine ingredients in whipper. Add nitrous. Enjoy. ◉

Share your nitrogen cavitation recipes, tips, and tricks at makezine.com/projects/nitrogen-cavitation

Water Bath Thermostat

For sous-vide and more — it controls both heating and cooling!

Written and photographed by Sean Michael Ragan

Time Required:
4 Hours
Cost:
$30–$45

Materials

- **STC-1000 temperature controller**
- **Junction box**
- **AC power outlet, Decora style**
- **AC power connector, IEC 320-C14**
- **Computer power cord**
- **Insulating binding post**
- **Crimp-on female disconnects: 22–18 AWG (1), 10–12 AWG (2)**
- **Wire**
- **Heat-shrink tubing**
- **Adhesive-backed label**

Tools

- **Screwdriver, #0 or #00 Phillips**
- **C-clamp**
- **Coping saw**
- **Drill and bits: ³⁄₁₆", countersink**
- **Label maker**
- **Pliers**
- **Printer**
- **Soldering iron and solder**
- **Wire cutter/stripper**

INSPIRED BY *COOKING FOR GEEKS* author Jeff Potter's quick DIY sous-vide cooking rig (makezine.com/pottersousvide), my plan was to just hack the temperature controller into an enclosure with an A/C outlet, the idea being that you could plug in any heater you wanted.

Looking around for cheap controllers, however, I happened across the STC-1000 on eBay for $25. It's not PID, but I was betting it would still be accurate enough for almost any practical purpose. And since the STC-1000 has both heating and cooling functions built-in, the logical next step seemed to be to split a single A/C outlet so that you could plug in a heater or a cooler — or both.

The STC-1000 will regulate at any tem-perature between the freezing and boiling points of water, which opens up potential applications in chemistry, aquaculture, zy-murgy, hydroponics, and cooking. Sous-vide eggs are a great place to start experiment-ing — I find an hour at 65°C (149°F) gives an absolutely perfect soft-boiled egg. ◉

For step-by-step instructions and images, visit makezine.com/projects/water-bath-thermostat

August Möbius
and his
Twisted Cylinder

Use the fascinating Möbius strip to make a spill-proof coffee cup carrier.

Written by William Gurstelle

Materials

- » **Acrylic plastic sheet,** 1/8" thick, 12"×12"
- » **Split rings, 1" diameter (2)**
- » **Barrel swivel, large** available where fishing tackle is sold
- » **Leather, 3½"×3½" piece**
- » **Soft leather, ½"×8" strip**
- » **Acrylic adhesive/solvent**
- » **Glue**

Tools

- » **Jigsaw** Use an acrylic cutting blade, or a carbide-tipped blade with fine teeth and little or no set.
- » **Hot air gun** for paint stripping
- » **Electric drill with 1/8" bit**
- » **Vise**
- » **Clamps, medium (2)**
- » **Sandpaper**
- » **Pop rivet gun and rivets, or stout needle and thread**
- » **Ruler and marking pencil**

WILLIAM GURSTELLE
is a contributing editor of *Make:* magazine. His new book, *Defending Your Castle: Build Catapults, Crossbows, Moats and More* is now available.

IN 1905, PHYSICIAN PAUL MÖBIUS PUBLISHED A PHOTO OF THE SKULL OF HIS LONG-DEAD GRANDFATHER, the great German mathematician August Ferdinand Möbius, juxtaposed with the skull of composer Ludwig von Beethoven. It's not clear how Paul obtained the photo of Grandpa Möbius' skull, much less that of Beethoven, but there they are, captured in black and white. This weird photo was made in the name of weird science. Paul Möbius was a neurologist of some distinction who dabbled in the bunk science of phrenology and evidently believed that certain skull shapes were associated with mathematical and musical aptitude.

While the younger Möbius' head bump research never panned out, he was right about one thing: August Möbius, professor of astronomy and observatory director at the University of Leipzig, was unquestionably a man of incredible mathematical ability. August Möbius' scientific contributions are enormous: He invented a new type of calculus, he advanced the fields of celestial mechanics and astronomy, and he provided many important insights into the study of geometry.

He is most famous today for his work in topology, the mathematical study of shape and form. Appreciating much of Möbius' work requires a familiarity with geometrical mechanics, polyhedral boundary theory, projective transformations, and other eye-glazing, no doubt important, mathematical concepts.

But there is one Möbian concept everyone can appreciate, and that is the wonderful Möbius strip. In 1858, Möbius took a band of thin, flat material, gave it a single twist and fastened the ends together. The Möbius strip was born. (To be completely accurate, it was independently discovered by mathematician Johannes Listing a few months before.)

Gunther Kirsch

In topological terms, a Möbius strip is a three-dimensional surface with only one side. It has some amazing physical properties. For example, drawn starting from the seam down the middle of the strip, a pencil line will meet back at the seam, but on the "other side."

If you cut a Möbius strip along this center line, you get not two separate strips, but rather one long strip with two complete axial twists. If you attach a Möbius strip to an object, say a bowling pin, and swing it around your head, the twisted strip resists kinking and curling, proving itself a superior attachment to a nontwisted one.

Let's exploit the properties of Dr. Möbius' strip to make a cup holder that's nearly spill-proof.

CAUTION:
WEAR GLOVES AND USE CARE TO AVOID BURNS. THE HOT AIR GUN AND PLASTIC BECOME EXTREMELY HOT!

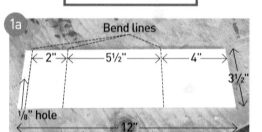

Bend lines
2" — 5½" — 4"
3½"
⅛" hole
12"

NOTE:
USE CLEAR WATER WITH YOUR COFFEE CUP CARRIER UNTIL YOU GET A FEEL FOR HOW IT HANDLES.

1. Make the cup holder

Cut a 3½"×12" strip of acrylic using the jigsaw (or table saw if you've got one). Mark the bend lines on the acrylic strip as shown in **Figure 1a**.

Place the acrylic strip in the vise and direct the hot air from the air gun at one of the bend lines. When the plastic softens, bend the piece 90°. Make the remaining bends (**Figure 1b**) so it takes the shape shown on page 92.

Using the bent plastic piece as a form, trace out and mark 2 side pieces approximately 6"×1½". Match the side contour of the main 12" piece as closely as possible (**Figure 1c**). Cut out the marked pieces.

Clamp the sides to the main piece and use acrylic cement to solvent-weld them (**Figure 1d**).

Drill a 1/8" hole in the top of the cup carrier assembly as shown in **Figure 1e**.

Glue the 3½" square leather pad to the bottom of the Coffee Cup Carrier assembly (**Figure 1f**).

2. Make the Möbius handle

Insert the 8" soft leather strip through one of the split rings. Turn the leather into a Möbius strip by matching up the ends and making a half twist, the top of one end on the bottom of the other end. Overlap the ends by about ¾" and fasten securely either with a strong needle and thread (use a thimble to push the needle through the leather) or with a couple of pop rivets (**Figure 2a**).

Insert the split ring with the Möbius strip into

one end of the barrel swivel.

Gently pry apart the other split ring and insert the other end of the barrel swivel. Now insert the split ring into the 1/8" hole on the cup holder assembly (**Figure 2b**). Your Möbius Strip Coffee Cup Carrier is ready.

3. Use it

Place a paper or plastic cup filled with water on the leather pad. Place one or two fingers through the leather handle. Begin by swinging the cup holder in small to medium circles. As you change the size of your swings — and even change directions — the water stays in the cup. It'll splash or spill if you jerk or whip the handle. But for most types of motion, the water will remain surprisingly steady in the cup.

How Does It Work?

To understand the Möbius Strip Coffee Cup Carrier, we must first consider why liquids spill. The physics behind sloshing coffee is complex. Not only is an intricate interplay of accelerations, torques, and forces at work, but the biomechanics of bipedal human motion come into play as well.

Researchers have found that the wave-like motion of coffee in a mug possesses a unique natural frequency that's related to the size of the mug (see journals.aps.org/pre/abstract/10.1103/PhysRevE.85.046117). Typical coffee cups produce oscillations that closely match the motion a person makes when walking. If you walk, coffee cup in hand, at a steady and even pace, there's no spillage. But even small irregularities in your gait cause breathtakingly complex accelerations of liquid that amplify the oscillations, leading to sloshing and ultimately, to stains on the carpet.

But add the Möbius Strip Coffee Cup Carrier with its kink-free handle, and the math becomes much simpler. Because the handle flexes easily, all forces acting on the coffee must act in line with the handle. As long as the handle stays taut, there are no lateral (that is, side to side) accelerations, and the forces acting on the coffee merely push it toward the bottom of the cup holder. ●

See more step-by-step photos and discuss coffee cup physics at makezine.com/mobius-strip-coffee-cup-carrier.

The Archivist:
A DIY Book Scanner

Build this rig with point-and-shoot cameras, skate bearings, and a Raspberry Pi to digitize your bookshelf.

Written by Daniel Reetz, Noah Bicknell, Johannes Baiter, and Matti Kariluoma

Time Required:
A Weekend
Cost:
$500–$1,000

Daniel Reetz/Eric Rojas

Materials

- » **Digital cameras (2)** Canon PowerShot A1400IS, $100 each
- » **SDHC memory cards, Class 10 (3)**
- » **Raspberry Pi B+ single-board computer** Maker Shed item #MKRPI5, makershed.com
- » **The Archivist Book Scanner Kit** about $700 from store.diybookscanner.org
- **— OR THE FOLLOWING PARTS:**
- » **USB Mini-B cables, 4' long (2)**
- » **USB Micro charger, at least 500mA**
- » **Outlet strip**
- » **Clamp lights, 5½" (3)**
- » **CFL bulbs, 14W (3)**
- » **Camera power supplies (2) (optional,** but recommended**)**
- » **Pulleys, 608ZZ type (3)** from store.diybookscanner.org, or use sheaves from National Hardware pulley #3213BC
- » **Framing screws, flat top, Phillips, ⁷⁄₁₆" (8)** black phosphate finish, Home Depot #100170183
- » **Skate bearings, 608ZZ type (20)**
- » **Threaded rod, M8, with 4 nuts** The kit uses some funky foreign bike skewers.
- » **Knobs with ¼-20 stud (2)** for camera rosettes, McMaster-Carr #7639K17
- » **Felt strips, self-adhesive, ½"×8"×⅛" thick (2)**
- » **Angle brackets, ⅝"×¾" (10)**
- » **Square tube or hardwood:** 1½", 28" lengths (2); 1", 28" lengths (2)
- » **Pipe or hardwood closet pole,** 1½" diameter, 30" lengths (2)
- » **Plywood, Baltic birch, 18mm, 4'×8' sheet**
- » **Bungee cords, 8mm×18" (2)**
- » **Adjustable bungee, 8mm×18"**
- » **Foamcore board, black,** 30"×40" sheets (2)
- » **Aluminum angle, 1/20" thick,** ¾" deep, 36" lengths (2)
- » **Paint, flat black**
- » **Various nuts, bolts, screws, and washers**

Tools

- » **Jigsaw or router with ¼" bit** A CNC router recommended, but it's doable with a hand router or jigsaw.
- » **Forstner bit, 22mm** for drilling bearing pockets
- » **Hex wrenches: 4mm and 6mm**
- » **Adjustable wrench**
- » **Screwdriver, #2 Phillips**
- » **Tape measure or ruler**
- » **Electrical tape**
- » **Double-sided mounting tape, high bond** such as 3M VHB or Scotch Outdoor Mounting Tape
- » **Tinsnips**

FIVE YEARS AGO WE BUILT OUR FIRST BOOK SCANNER FROM SALVAGE AND SCRAP. Book digitization was the domain of giants — Microsoft and Google. Commercial book scanners cost as much as a small car. Unless you chose to destroy your books in sheet-feed or flatbed scanners, there was no safe and affordable way to preserve the contents of your bookshelf on your e-reader.

Collectively, we tried to fix that. Over 2,000 people contributed more than 350 designs and thousands of lines of code at diybookscanner.org.

The result is the Archivist — the VW Beetle of book scanners — cheap, durable, and tremendously effective. It's open source and made with the simplest materials possible, like plywood, bungees, and skateboard bearings. As fast as you can turn the pages, the Archivist photographs them automatically and creates a zip file of the entire book, for conversion to the e-book format of your choice.

HERE'S HOW DIY BOOK SCANNING WORKS:

- » **Lighting** — Cheap cameras need a lot of light, and post-processing is easier if the light is bright and uniform.
- » **Camera support** — Cameras need to be placed opposite the page. Post-processing is vastly easier if the cameras don't move.
- » **Platen** — The "page flattener." It's a V-shaped construction of glass or acrylic. Flat pages are easier to process — in fact they're usable with no post-processing. (Good input equals good output!)
- » **Cradle** — Holds the book, gently, and accommodates the spine of the book.
- » **Base** — Holds all the other stuff together.
- » **Electronics** — Many of us build a simple system to trigger our cameras electronically using Stereo Data Maker software. The new Archivist system has an onboard Raspberry Pi running software called Spreads to operate two Canon point-and-shoot cameras (updated with Canon Hacks Development Kit firmware) and then sends the zipped images to your computer. Spreads even has a web interface so you can control your book scanning from any smartphone or tablet. ◢

Build your own Archivist book scanner with complete step-by-step photos, instructions, and downloads at makezine.com/projects/diy-book-scanner.

Mini Spin Art Machine

Written by Riley Mullen ■ Illustrations by Julie West

1

STICK IT

BOX INSIDE

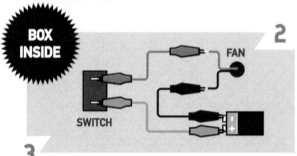

FAN

SWITCH

2

3

NEAT!

THIS PROJECT CAME FROM MY LOVE OF TAKING APART BROKEN ELECTRONICS AND EXPERIMENTING WITH THE COMPONENTS. I had an old computer fan, a 9-volt battery, pens, and the gift of boredom. Next thing I knew, I was creating interesting art! Turning electronic "junk" into unexpected fun is really satisfying.

1. Mount the fan and switch

Attach the fan to the box top with double-stick foam tape. »» To mount the switch, make a small hole in the box. »» Remove the mounting nut from the switch, push the switch lever through the hole from the inside of the box, and replace the mounting nut to secure it in place.

2. Wire the fan, switch, and battery

Make another small hole to push the fan wires through to the inside of the box. »» Using one alligator clip lead, connect the fan's black wire to the negative (larger) terminal on the battery. »» Using another alligator clip lead, connect the fan's red wire to one of the switch terminals. »» Use the last alligator clip lead to connect the unused switch terminal to the positive (smaller) terminal on the battery.

3. Add paper and color

Cut paper circles smaller than the diameter of your fan. »» Attach a paper circle to the center of the fan, using a loop of masking tape on the back of the paper. »» Turn the fan on and gently apply a colored marker to the spinning paper. Enjoy making colorful designs! ◐

Watch mini spin art being made: makezine.com/projects/spin-art-machine

RILEY MULLEN is an 11-year-old maker who is endlessly fascinated with electronic components, physics, and engineering. He enjoys reading, tinkering, creating things with household stuff, and hanging out with other makers at his local makerspace.

You will need:

- »» **Computer fan, 5V–12V**
- »» **Battery, 9V**
- »» **Toggle switch, two-terminal** RadioShack #275-0022
- »» **Alligator clip leads (3)**
- »» **Cardboard box**
- »» **Double-stick foam tape**
- »» **Masking tape**
- »» **Scissors**
- »» **Paper**
- »» **Colored markers**

Thomas Jefferson:
Maker-in-Chief

A peek in the president's home reveals his penchant for making.

Written and photographed by Forrest M. Mims III ■ Illustrated by Nate Van Dyke

THERE IS NO "MAKER IN CHIEF" IN THE U.S. CONSTITUTION, BUT AT LEAST ONE PRESIDENT, THOMAS JEFFERSON, CERTAINLY DESERVES THAT TITLE.

Jefferson is most famous for drafting the Declaration of Independence and serving as the third president of the United States. It's less well known that he was also an experienced architect, surveyor, locksmith, and amateur scientist. And he was an innovator who made improvements in the design of clocks, instruments, and the polygraph copying machines that duplicated his letters as he wrote them.

Visitors to Monticello, Jefferson's mountaintop home near Charlottesville, Virginia, can quickly become acquainted with Jefferson's maker side. Walking toward the front porch, you can see the entire house and its famous dome — all designed by Jefferson. "Architecture is my delight," he told a visitor, "and putting up and pulling down one of my favorite amusements."

The Wind Vane

Another clue at Monticello is the large weather vane atop the front porch. The vane's shaft extends through the roof of the porch to the ceiling, where it is attached to a pointer that indicates the direction of the wind on a compass rose.

FORREST M. MIMS III
(forrestmims.org), an amateur scientist and Rolex Award winner, was named by *Discover* magazine as one of the "50 Best Brains in Science." His books have sold more than 7 million copies.

Jefferson and his family could check the wind direction simply by looking through one of the nearby windows.

The Great Clock

Above the front entrance is another Jefferson innovation, his double-sided Great Clock. The clock's movement and its dial indicating hours, minutes, and seconds are installed in a wood frame mounted inside the house, over the entrance door. A second dial that indicates the hour is mounted outside above the door.

The Great Clock was built to Jefferson's specifications by Philadelphia clockmaker Leslie & Price in 1793 and installed at Monticello in 1804. Its mechanism is powered by six cannonball-like weights suspended by a rope and pulley along the corner of the right side of the entrance hall. The days Sunday through Thursday are marked on the wall adjacent to the weights; as the weights descend, the topmost weight indicates the day of the week. Eventually, the weights drop through a circular hole in the floor to the basement, where Friday and Saturday are marked on the wall.

A Chinese gong indicates each hour. The gong's striking mechanism is powered by a set of eight weights suspended by a rope and pulley along the corner of the left side of the entrance door.

Each Sunday, Jefferson climbed a folding ladder to wind the clock. When not in use, one side of the ladder was pushed up to merge it with the other side. The folding ladder was built in Monticello's wood shop, known as the joinery.

Jefferson's obsession with accurate timekeeping was closely related to his interest in astronomy and telescopes, several of which he purchased over the years. He described astronomy as "the most sublime of all the sciences."

Astronomy also served a practical purpose, for Jefferson was committed to measuring the geographic coordinates of Monticello as accurately as possible. With help from President James Madison — who lived nearby and often spent time at Monticello — Jefferson measured the timing of the annular solar eclipse of 1811. The times he and Madison measured were used by amateur eclipse expert William Lambert to calculate the longitude of Monticello as –78.50°, which is very close to Google Earth's –78.45°.

The Spherical Sundial

During Jefferson's time, the accuracy of clocks depended on a high-resolution sundial. Jefferson had probably seen spherical sundials in Europe, and he designed one for Monticello that was a 10½-inch sphere made from carefully inscribed locust wood fitted with a movable sundial blade. The spherical sundial was built in the joinery to his exact specifications. The fate of the original is unknown, but the Thomas Jefferson Foundation commissioned a replica based on Jefferson's detailed design. It is now installed at Monticello near where the original stood.

The Open-Source Plow

Perhaps the most down-to-earth of Jefferson's innovations was his design of an improved moldboard for plows. The moldboard is the section of the plow that lifts and turns the soil cut by the plow's leading edge. Jefferson claimed his moldboard was more efficient than previous designs. He did not apply for patent protection for his design. Instead, he sent models of his moldboard design to others along with details about its design and construction.

Jefferson's Tools

While Jefferson was the architect of Monticello and the designer of many of its clocks, instruments, and mechanical contrivances, he did not personally build all of these things. But this does not disqualify him from being a hands-on maker. Jefferson's papers and correspondence mention his tools and that he took some of them to Paris when he served as U.S. minister to France from 1785 to 1789.

Several of Jefferson's contemporaries wrote about his many personal tools, including Isaac Granger, who was born into slavery at Monticello and whom Jefferson took to Philadelphia, where he learned to be a tinsmith. When interviewed by the Rev. Charles Campbell in 1847, Granger recalled that, "My Old Master was neat a hand as ever you see to make keys and locks and small chains, iron, and brass. He kept all kind of blacksmith and carpenter tools in a great case with shelves to it in his library."

Presidential Props

While addressing a dinner to honor Nobel Prize winners from the Western Hemisphere on April 23, 1962, President John Kennedy said, "I think this is the most extraordinary collection of talent, of human knowledge, that has ever been gathered together at the White House, with the possible exception of when Thomas Jefferson dined alone." If President Jefferson were alive today, we'd like to think he'd be among the many *Make:* magazine readers. ●

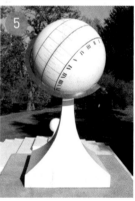

1. The west side of Jefferson's Monticello home.

2. The wind vane over the front portico of Monticello.

3. The wind direction is indicated on this compass rose on the ceiling of the front portico.

4. The outside dial of the Great Clock is over the entrance to the front portico.

5. A replica of Jefferson's spherical sundial.

Solder a "Draailampje"

Flip lights! Turn them upside down, and they're off.
Turn them back up, and they're on. Written and photographed by David Bakker

**Time Required:
15-30 Minutes
Cost:
$2-$29**

DAVID BAKKER
is a maker. He organizes
the Groningen Mini
Maker Faire (Netherlands)
and trains people to
speed-read and time
manage.

Materials
» Jar, small
» LED, 10mm, bright
» Tilt/angle switch
» Battery, CR2032
» Battery holder, CR2032
» Glue or foam tape

**THIS CUTE LITTLE SELF-CONTAINED LAMP
WORKS LIKE A CHARM** — set it right side up to
turn the light on, flip it upside down to turn it off
again. I call it the *draailampje* — that's Dutch for
"flip light."

While most jars will work, I find a small,
hexagonal jar to be particularly pretty. My first
containers were mini jam jars from a hotel. Now I
use 47mm hexagonal jars, which I buy in bulk.

1. Solder
Fold the legs of the battery holder flat to the
outside, then solder the negative (–) lead from the
LED to the – lead from the battery holder.

Solder one leg of the tilt switch to the positive
(+) lead of the battery holder and the other leg to
the + lead of the LED (some bending required).

2. Add the battery
Insert the battery in its holder, and glue the holder
to the jar lid.

3. Use your draailampje
Next to the bed, at the campsite, in the rain — this
handy lamp has many uses. And the light will

remain bright for 5 days and nights of continuous
use — just replace the inexpensive battery when it
starts to dim.

Going Further
Why not upgrade your draailampje? Try adding
a slow-fading multicolor LED, or making it solar
powered or rechargeable.

Making a flip light is easy and fun — great
for nontechnical people or first-time solderers,
it makes for a perfect workshop or classroom
activity. ◔

For detailed step-by-step build instructions,
visit makezine.com/projects/draailampje

Glowing, Dancing Oobleck

Written by the *Make:* editors ■ Illustrations by Julie West

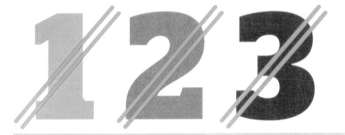

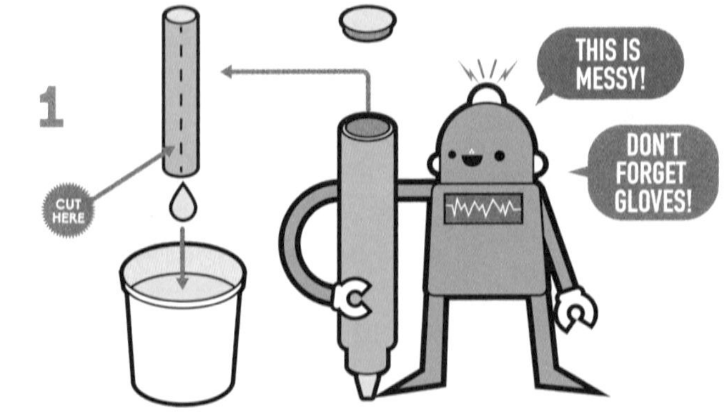
1
CUT HERE
THIS IS MESSY!
DON'T FORGET GLOVES!

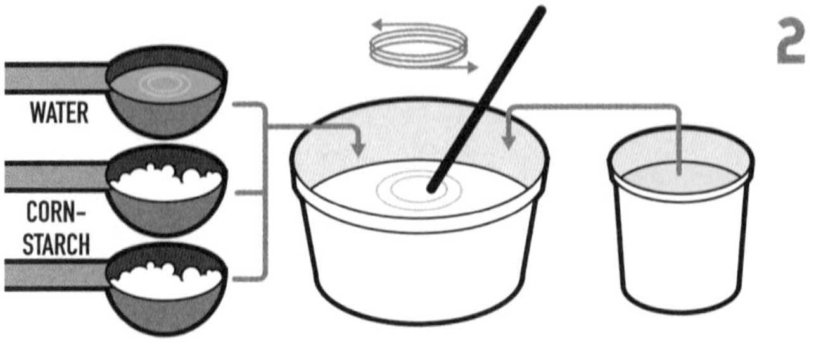
2
WATER
CORN-STARCH

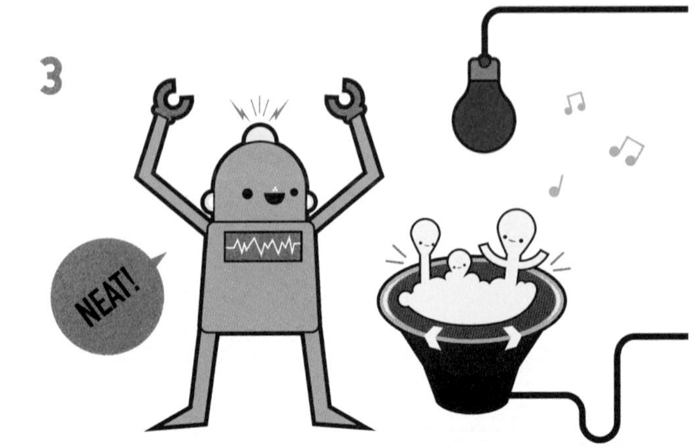
3
NEAT!

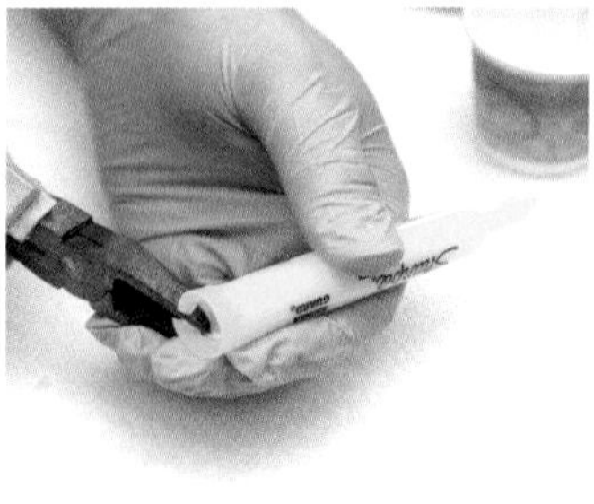

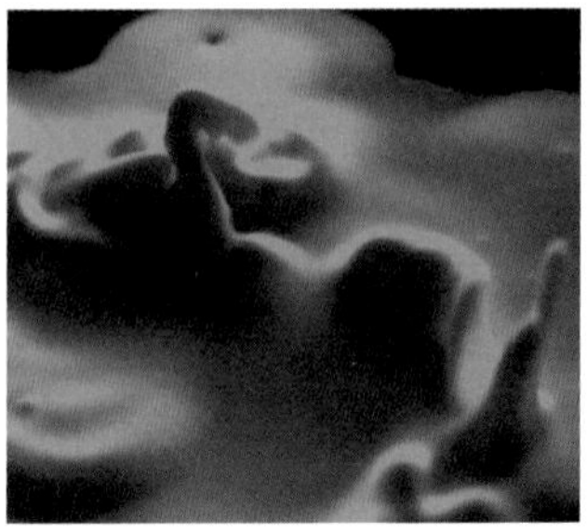

***OOBLECK* IS A DELIGHTFUL NON-NEWTONIAN FLUID** that exists as a liquid under low-stress situations, but becomes more viscous when agitated, either by hands or by vibrations. Named after a fictional green precipitation in a Dr. Seuss book, the substance is fun on its own, but you can add more awesome by making it glow (with highlighter dye, under a black light) and making it perform (by placing it on a speaker).

1. Extract highlighter dye

Cut open the bottom of the highlighter using the cutting pliers. » Put on the rubber gloves, remove the ink cartridge, and cut it open straight down the middle, from top to bottom, using the box cutter. » Remove and squeeze the spongy material from the top down to drain all the ink out of it.

2. Mix and pour

Make the oobleck by mixing the cornstarch with water in a ratio of 2 parts cornstarch, 1 part water. » Add the highlighter dye. Adjust the ratio until it's runny but tough to mix. » Cover the subwoofer tightly with plastic wrap, using tape to anchor it to the bottom. The tighter the wrap, the better the results. » Pour the oobleck onto the plastic wrap.

3. Play tones

Hook up the subwoofer to your computer and experiment with playing different frequencies to find the sweet spot. We noticed interesting activity in the 20Hz–60Hz range. » Turn off the lights and turn on your black light. » Watch the effects on the oobleck as it takes on lifelike qualities, assembling itself into interesting patterns and abstract, creature-like forms.

Watch the glowing oobleck in action:
makezine.com/projects/oobleck

You will need:

- **Cornstarch**
- **Water**
- **Subwoofer**
- **Plastic wrap**
- **Tape**
- **Computer** or other device to generate different frequencies
- **Black light**
- **Highlighter pen, yellow, nontoxic**
- **Rubber gloves**
- **Box cutter**
- **Cutting pliers or tinsnips**
- **Bowl**
- **Spoon**

Creating Cosmic Couture with Mylar

Time Required: 1 Hour Cost: $0–$5

Written by Michelle Hlubinka

MICHELLE HLUBINKA is director of custom programs for Maker Media, overseeing outreach and programming to help schools, kids, and families get into making, including her own two sparkly young boys (Ion is pictured above, left).

You will need:

» **Mylar "space blankets"** A lot of bang for the buck; the stuff is cheap and surprisingly squishable for easy storage.

» **Scissors**

» **Invisible tape** aka Scotch tape; easy to remove

» **Flexible aluminum duct, 3" (optional)** for robot and astronaut arms

» **Cardboard boxes (optional)** for robot bodies

» **Decorations (optional)** stickers, duct tape, and such

AT 2012'S EAST BAY MINI MAKER FAIRE AT PARK DAY SCHOOL IN OAKLAND, one room became resplendent. Silvery, shimmery sheets of mylar draped over its doors and all its surfaces. Kids and adults alike crafted the mylar into hats, gowns, spacesuits, and robot costumes. Stepping into the room transported me to some campy sci-fi classic (*Barbarella? Saturn 3?*). Later, I spotted brilliant fashionistas bouncing around the event wearing the products of their intergalactic sculpting.

For my eldest's fifth birthday party, we set up a sparkly "Cosmic Couture" fashion studio inspired by the mylar room. (It was a standout among several space-themed activities, including compressed air rocket launches in the front yard and a moon bounce in the backyard.) I bought 20 space blankets for about a buck each and watched the mylar metamorphosis unfold.

To make a costume, start by cutting a hole in the middle for your head, and snipping and tucking in loose pieces and excess with tape until you get a base shape. Tape long tubes for sleeves. Or use flexible aluminum duct: Snip to length with a wire cutter and old scissors, then cover the ends with silver duct tape to stave off scratches. We found 3" duct worked well as robot arms for our 3-year-old son's costume. Add flourishes with other pieces. We recommend using tape rather than hot glue, as mylar is very flammable.

We've gotten a lot of mileage out of the mylar. We reused the pieces from the Cosmic Couture activity to create a 15-foot-long "wormhole" tunnel for our space-themed Apollo-ween display last October. Mylar can also be used to project fantastic light displays on a ceiling. It's both translucent and reflective.

Mylar's magnificent. To get shiny, crafty kids often use aluminum foil, but mylar comes in much larger and lighter sheets, and it's more robust than that ol' kitchen standby. ●

See more photos and share your mylar projects at makezine.com/projects/ cosmic-couture

Jeffrey Braverman

DIY BUILDS:
CREATIVE KITS

SOLARBOTICS
www.solarbotics.com

Powered by free solar energy, this machine kit keeps on rolling!

Toy Inventor's Notebook

MAGIC CAT PROJECTOR

Invented and drawn by Bob Knetzger

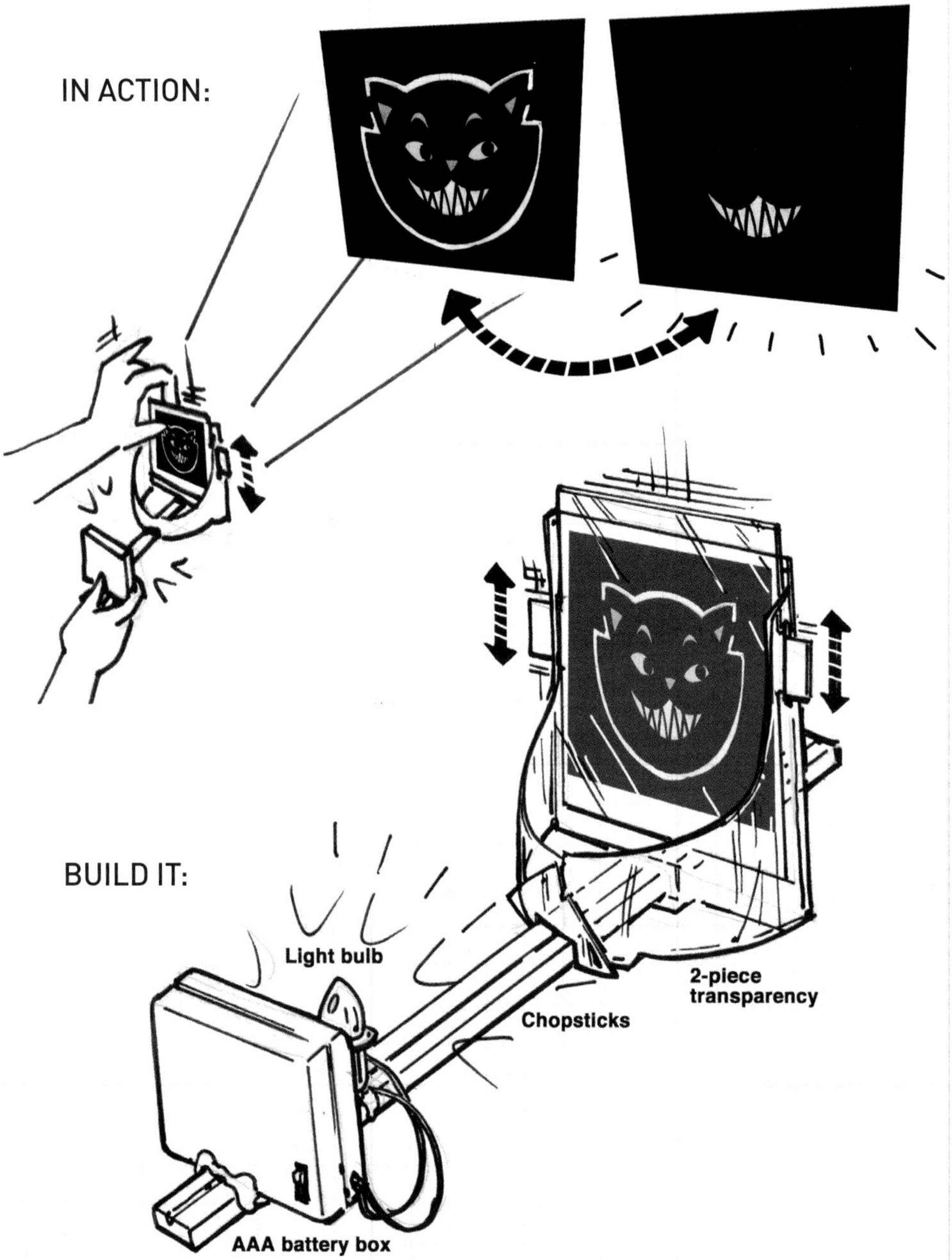

IN ACTION:

BUILD IT:

Light bulb

2-piece transparency

Chopsticks

AAA battery box

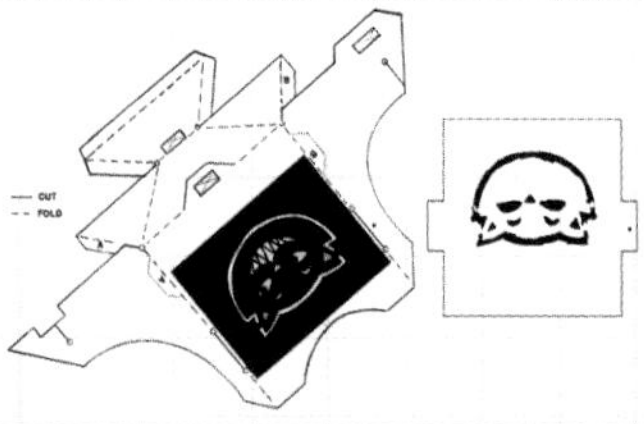

Get detailed folding instructions, download the template, and see the Magic Cat Projector in action at makezine.com/magic-cat-projector.

MAKE THIS FUN MINI PROJECTOR JUST IN TIME FOR HALLOWEEN. It throws a shadow image of a black cat — which then disappears, leaving only his spooky grin!

No lenses or focusing needed. Just go to makezine.com/magic-cat-projector to download the image file. Print it out full size on a single sheet of transparency film. Use a hobby knife to cut out the 2 shapes. Carefully cut the slits as shown and use a small-size (⅛" hole) paper punch to make the little circles to provide stress relief at the inside corners.

Score on the dashed fold lines by tracing over them with a ballpoint pen and straightedge. Then fold as shown. The sides fold forward and interlock. Fold the bottom backward and bend the 2 wings up and tape together at A and B. Thread an unbroken pair of takeout chopsticks through the square holes. Insert the tabs of the second shape into the slits, matching the asterisk on the first shape. It should freely slide up and down in the slits.

Solder the 2 leads from a four-AAA battery pack to the contacts of a PR13 flashlight bulb. The battery box I used had a built in on/off switch. If yours doesn't, just take the batteries out to turn it off. Hot-glue the battery box to the other end of the chopsticks. Hot-glue the light bulb to the chopsticks and to the backside of the battery box. Dim the lights and slide the film up and down to make the cat disappear and reappear!

Feeling ambitious? Take the project to the next level by adding a small motor and cam to automate the effect into a free-running display. Or replace the flimsy film with some laser-cut fiberboard (first add some bridges and a frame to the isolated image "islands" so it all holds together.)

Happy Halloween! ◆

TOOLBOX

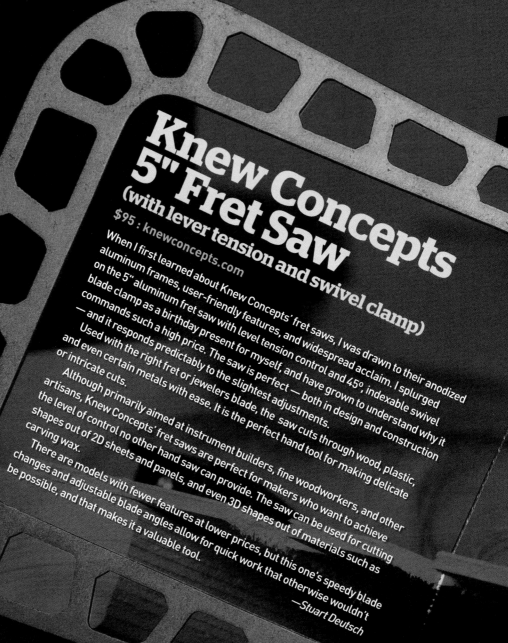

Knew Concepts 5" Fret Saw
(with lever tension and swivel clamp)

$95 : knewconcepts.com

When I first learned about Knew Concepts' fret saws, I was drawn to their anodized aluminum frames, user-friendly features, and widespread acclaim. I splurged on the 5" aluminum fret saw with level tension control and 45°, indexable swivel blade clamp as a birthday present for myself, and have grown to understand why it commands such a high price. The saw is perfect — both in design and construction — and it responds predictably to the slightest adjustments.

Used with the right fret or jewelers blade, the saw cuts through wood, plastic, and even certain metals with ease. It is the perfect hand tool for making delicate or intricate cuts.

Although primarily aimed at instrument builders, fine woodworkers, and other artisans, Knew Concepts' fret saws are perfect for makers who want to achieve the level of control no other hand saw can provide. The saw can be used for cutting shapes out of 2D sheets and panels, and even 3D shapes out of materials such as carving wax.

There are models with fewer features at lower prices, but this one's speedy blade changes and adjustable blade angles allow for quick work that otherwise wouldn't be possible, and that makes it a valuable tool.

—Stuart Deutsch

MILWAUKEE STEP DRILL BITS

$20+ : milwaukeetool.com

Step drill bits are multilevel bits that can be used to bore many different sizes of holes without having to change bits. They are especially useful for drilling holes in sheet metal and plastic panels. It's difficult for companies to innovate with step drill bits, as there isn't a lot of room for improvement. However, these new bits seem to be an exception.

Flats ground into the shank of each bit are firmly gripped against the jaws of a drill chuck, ensuring secure slip-free drilling through tough materials. The tip of each bit has a relief carved under the cutting edges, which helps to keep it from wandering when starting a hole, even on curved surfaces like metal piping. My favorite feature is the silvered lines across the most common steps, such as ¼", ⅜", and ½", which are still visible when the bit is spinning. This helps me know when holes are drilled to the desired diameters.

There are cheaper brands of step drill bits, but it is clear that these Milwaukee bits are of a higher caliber.

—*Dan Spangler*

3M VHB PERMANENT DOUBLE-SIDED FOAM TAPE

Price varies : 3M.com

3M's VHB tape is a super versatile, double-sided, semi-permanent, durable, all-weather, all-conditions tape that simplifies things when you want to attach parts to other surfaces with minimal effort. The tape ensures that attachments stay secure and exactly as intended.

3M's engineering experts designed this tape as an alternative to bolts, rivets, screws, and welds to reduce the level of permanency and work required for quick attachments and installations. It is important to point out that, once applied and set up, it can be quite a chore to remove the tape.

The main advantage is the plethora of applications it is suitable for. Makers can use this tape anywhere they don't want to drill holes, such as to attach LCD or glass panels to the inside of bezels and project enclosures. This tape is also preinstalled on GoPro video camera mounting accessories, which should give you a good idea of its trustworthiness. —*Enrique Depola*

DREMEL ULTRA-SAW US40

$129 : dremel.com

The Dremel US40 Ultra-Saw seems like one of those niche tools that a lot of makers might overlook, but only at first. As soon I started using it I thought to myself, "Why hadn't I gotten one of these sooner?"

The saw offers the perfect balance between a circular saw and an angle grinder or cut-off tool. Its smaller form factor makes it feel more controllable, and possibly even safer. Unlike larger circular saws, which can sometimes be a little terrifying to use, the Ultra-Saw is powerful, yet with a very tame feel.

In addition to its woodcutting abilities, the saw triples as both a metal cutter and a surface sander. Anyone who works in a cramped and cluttered garage will especially appreciate the multifunctional versatility of this one tool. —*DS*

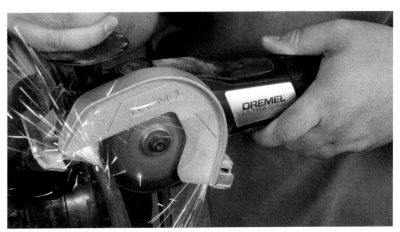

TOOLBOX

YOST COMPACT BENCH VISE (RIA-4)

$117 : yostvises.com

The Yost Compact Bench Vise features 4" jaws that open 2 ¼" wide, a swivel base with 360-degree range, and a flat anvil spot behind the jaws for pounding on. It's a small bench vise that's a good fit for small jobs, and its compact footprint makes it a good match for smaller workspaces.

The small vise pairs well with Dremel-sized projects. But despite its cozy size, the vise feels strong and solid. The jaws open smoothly and tighten squarely, and the vise turns well and locks easily on its swivel base. I trusted it to hold work against a 23,000-RPM rotary tool, and it did the job beautifully.

If you work with your hands, you probably already know that workholding can be crucial to achieving good results. There are times when you can hold a workpiece in your hands, but a good vise will not only make the work stay put, it'll protect your fingers and hands as well. Any shop worth its salt will have at least one solid vise. If small projects are your deal, or if bench space is at a premium, this may be the one for you.

—Sam Freeman

TEKTON PLIERS

$8+ : tektontools.com

Whether I'm tuning up my bike or putting together a new project, I always seem to be reaching for a pair of pliers. These new Tekton pliers are ruggedly built in the USA and have incredibly smooth, yet reliable, tongue and groove joints. The handles are ergonomically designed and provide plenty of comfort when applying large amounts of torque. The best part is the powerful set of teeth that prevent any slipping. All combined, these features have made Tekton to my new go-to brand for pliers.

—DS

THROUGH-HOLE NEOPIXELS

$4.95 for 5 : adafruit.com

NeoPixels, Adafruit's WS2812 LED products, are individually addressable RGB LEDs that can be connected together and controlled by a single microcontroller data pin to generate almost any glowing color imaginable. Previously, individual NeoPixels were only available in surface-mount (SMD) and breadboard-friendly packages, but they are now available in through-hole packages in both 5mm and 8mm sizes.

Because these NeoPixels have standard LED leads, it has never been easier to use individual NeoPixels in your projects. The 5mm and 8mm packages make these LEDs perfect as indicators, but they can also be used for creative projects or anywhere you might want or need a larger, customizable light source.

—SD

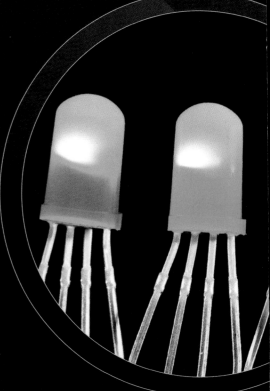

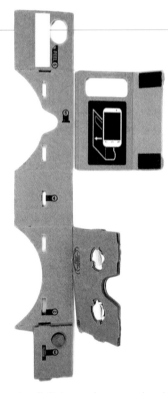

GOOGLE CARDBOARD

Free design files,
$25 kit : g.co/cardboard

There's no doubt that Virtual Reality is making a comeback. Now Google is jumping into the fray with Cardboard, a no-frills, assemble-it-yourself enclosure for your phone that adapts it into a head-mounted display. Materials in the build can be sourced easily and include cardboard (naturally), lenses, magnets, velcro, and a rubber band. By creating an inexpensive tool for VR, Google hopes "to encourage developers to build the next generation of immersive digital experiences and make them available to everyone."

In addition to offering the plans to make the head-mounted enclosure, Google released a free experimental VR toolkit for Android so that you can create your own interactive virtual reality experiences.
—Matt Richardson

FONA MINI CELLULAR GSM BREAKOUT

$40 with uFL Antenna Connector, $45 with SMA Antenna Connector : adafruit.com

To help you create wireless devices that can go far beyond the boundaries of your wi-fi network, Adafruit introduced FONA, their brand new GSM breakout board. The tiny board is capable of connecting to cellular networks for voice, text, and data communications, enabling your project to roam much of the world and still phone home.

Supply your own activated SIM card and use your development board of choice to communicate with FONA via serial with AT commands. FONA's extra features, such as battery charging circuitry, headphone jack, and indicator LEDs make working with cellular communication easier than ever.
—MR

MeArm

Free design files, £24.99 kit ($43) : phenoptix.com

Designed by Benjamin Gray and Jack Howard at phenoptix, MeArm is a low-cost, open-source robotic arm. Most of the arm assembly is made from laser-cut acrylic held together with a few screws and nuts. The joints are actuated with hobby servos, which can be controlled by any number of development boards. They have provided code examples to help Raspberry Pi and Arduino enthusiasts get started.

You can order a kit directly from phenoptix or download their design files and make your own. Whichever way you do it, MeArm is an inexpensive and fast way to get your 'bot to grab.
—MR

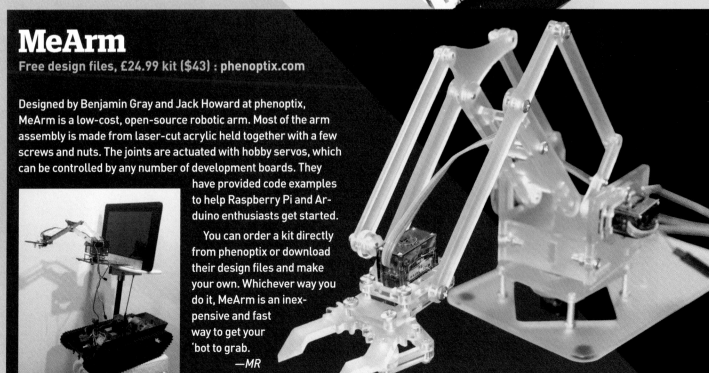

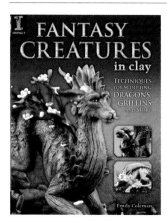

FANTASY CREATURES IN CLAY

By Emily Coleman
$25 : Impact

As a sculptor, I know how mystifying it is for a newbie to navigate materials, tools, and techniques while discovering your specific style and focus. The biggest obstacle is lack of a single source of modern techniques and materials that is visual in its delivery.

Emily Coleman's *Fantasy Creatures in Clay* hits the mark on so many levels that I urge anyone starting off to purchase it. Don't be turned off by the cute dragon on the cover — Emily's love of bubblegum fantasy creatures happens to be her signature style. Dig in and you will find straightforward tutorials on making sturdy armatures, building forms, detailing in a variety of styles (feathers, scales, etc.), and painting.

When you master this book, move on to my all-time favorite, *Pop Sculpture: How to Create Action Figures and Collectible Statues*, a natural progression of techniques, materials, molding and casting.

—Jason Babler

THE MARTIAN: A NOVEL

By Andy Weir
$24 : Broadway Books

The Martian is a celebration of the eternal problem solvers. A man stranded in the most hostile environment possible needs to figure out how to survive with the tools on hand. Andy Weir's attention to detail as well as his unique sense of humor turn what is essentially a really long equation into a gripping story of life and death.

Andy paints the picture so clearly that you can see this individual as he scuttles around the surface of Mars, jumping from one cobbled-together project to another in an attempt to get back home. *The Martian* is a true maker's tale and an absolute must read.

—Caleb Kraft

3D MODELING AND PRINTING WITH TINKERCAD: CREATE AND PRINT YOUR OWN 3D MODELS

By James Floyd Kelly
$30 : Que Publishing

When I was writing my book, *Making Simple Robots*, I tried several computer-aided design programs to find one readers could use to make 3D-printed parts. In the end I went with Tinkercad, Autodesk's free online CAD application for beginners and kids. I love the drag-and-drop mechanics that start off with solid objects you can squash, stretch, and even carve using solid-shaped "holes." But Tinkercad's online support is less robust; you're left combing through blog posts and forums looking for help with undefined tools and functions. James Floyd Kelly's *3D Modeling and Printing with Tinkercad* fills that void, using fun projects to explain the application's features. Kelly even covers 3D modeling basics and ways to take a project from screen to printer. Helpful tips (like how to instantly rotate an object 45°) and full-color photos make this a great guide to keep on hand.

—Kathy Ceceri

New From MAKER MEDIA
MAKE: ROCKETS

By Mike Westerfield
$45 : Maker Media

Are you ready to become a rocket scientist? It's easier than you think!

Hobby rockets commonly break the 100,000-foot altitude barrier. With this book, you'll build high-performance rockets that can hit 500mph and fly half a mile into the sky, and experience the thrill of seeing a glider roar up and come safely back down.

Using the full-color illustrations and helpful photos, elementary school students can build projects in this book with the help of parents and teachers. High school students may need just a little guidance..

Starting with the basics of rocket propulsion, you'll begin making rockets crafted from stuff you probably already have around the house. Best of all, you'll learn how to fly your rocket safely, recover it sanely, and follow the rules for hobby rocketry.

—Gretchen Giles

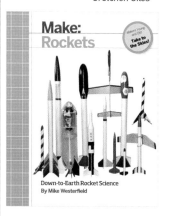

Make: Rockets
Down-to-Earth Rocket Science
By Mike Westerfield

TINKER WITH:
WEARABLES

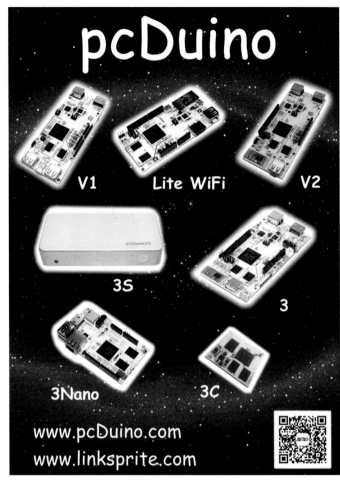
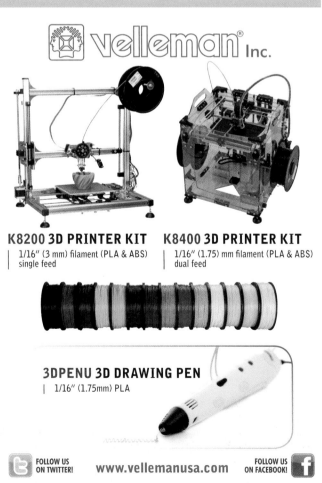

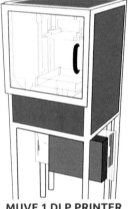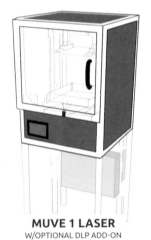

"La Princesse"

Written by James Burke

HOW CAN I LOVE SOMETHING WITH EIGHT EYES AND A BLOODLESS HEART? I just don't think there's a newspaper large enough to smash this beast. Not that I would, but spiders never made me feel very comfortable. They haunt every crack and corner in your house and live with some salesman's promise to "rid your home of all other bugs" in exchange for the occasional fear of accidentally climbing into your slumbering bedtime mouth. That's not entirely factual, and neither is the fear factor from a 50-foot arachnid that is always within your field of view whenever it chooses to terrorize a city.

200,000 attendees (or victims?) of La Princesse witnessed five days of nigh-Lovecraftian horror back in 2008 during Liverpool's Capital of Culture event and it has remained in my nightmares ever since. I can't fathom an evening without seeing that mechanical metatarsus, or the sunken forlorn eyes of its diabolical builders, the fire-breathing operators from French production company La Machine.

The crowd was fanatically attracted by the musical pheromones that spewed from the machines' spinnerets. When all was settled the freak nestled into its cocoon and the gasoline engines, compressors, and fireworks went silent and she was never seen again.

At least I want to believe that. Let me have that moment of rest. ⊘